ECOLOGY AND EVOLUTIONARY BIOLOGY

ECOLOGY AND EVOLUTIONARY BIOLOGY

A Round Table on Research

EDITED BY

George W. Salt

The University of Chicago Press
CHICAGO AND LONDON

The articles in this volume originally appeared in the November 1983 issue of *The American Naturalist* Volume 122, number 5.

The University of Chicago Press, Chicago 60637
The University of Chicago Press, Ltd., London
91 90 89 88 87 96 85   5 4 3 2 1

© 1984 by The University of Chicago
All rights reserved.   Published 1984
Printed in the United States of America

Library of Congress Cataloging in Publication Data
Main entry under title:

Ecology and evolutionary biology.

   Also published as The American naturalist, v. 122, no. 5, November 1983.
   Includes bibliographical references and index.
   1. Ecology—Addresses, essays, lectures. 2. Evolution—Addresses, essays, lectures. 3. Biotic communities—Addresses, essays, lectures. I. Salt, George W.
QH541.145.E259   1984   574.5   83-24088
ISBN 0-226-73443-9

CONTENTS

Introduction. *George W. Salt*	1
Competition and theory in community ecology. *Jonathan Roughgarden*	3
On hypothesis testing in ecology and evolution. *James F. Quinn and Arthur E. Dunham*	22
Detecting community-wide patterns: estimating power strengthens statistical inference. *Catherine A. Toft and Patrick J. Shea*	38
Competition theory, hypothesis testing, and other community ecological buzzwords. *Daniel Simberloff*	46
Natural variability and the manifold mechanisms of ecological communities. *Donald R. Strong, Jr.*	56
On the prevalence and relative importance of interspecific competition: evidence from field experiments. *Joseph H. Connell*	81
Roles: their limits and responsibilities in ecological and evolutionary research. *George W. Salt*	117
Index	127

Introduction

In recent years, two powerful intellectual currents in ecological and evolutionary thinking have met, not head on, but at an angle. The turbulence resulting from this confluence has generated papers, symposia, seminars, and informal discussions.

One current represents the efforts of those individuals whose goal is the removal of the lamentable displacement between ecological and evolutionary thinking. Because the phenomenon of competition is central to the theory of natural selection and has long been a major topic in ecology, it is natural that this process should be the point at which they seek to bridge the two disciplines.

The other current is generated by those who take their point of view from the writings of the philosopher Karl Popper and specifically from his insistence that a hypothesis (or model) cannot be considered scientific unless it is framed in such a way as to be subject to disproof. These individuals criticize the writings of the "bridgers" not so much for their goals as for the methods they employ.

By and large, the exchanges have taken place at long range, much like artillery fire between batteries hidden behind hills. An article appears and its answer follows weeks or months later, sometimes in a different journal or symposium volume.

When the editorial office of *The American Naturalist* received a manuscript from Dr. Jonathan Roughgarden, who sought to answer critics of his writings, it was apparent that there would ensue another round of exchanges with the usual time lags between. Each author named in his manuscript would undoubtedly write a reply and expect that it should be published, and, equally undoubtedly, Dr. Roughgarden would provide a rebuttal, and so on. It appeared that it would be more fruitful for the readers of the journal if all the exchanges took place simultaneously and the authors' various position statements were presented together.

Although it meant a considerable delay in the appearance of his article, Dr. Roughgarden agreed to this arrangement, and statements were solicited from Drs. Simberloff, Strong, and Connell. Independent manuscripts submitted to *The American Naturalist* by Drs. Toft and Shea and by Drs. Quinn and Dunham were added to the initial papers to provide a form of round table. All manuscripts were circulated to all participants, and queries and objections were passed back and forth so that all participants had the opportunity to comment on the statements made by all other individuals in advance of publication.

This process of direct exchange did not institute a serene calm in place of the former turbulence, but it did serve to generate a more analytical and dispassionate statement of views than had been available before. As such, it is valuable. Furthermore, despite the superficial excitement which controversy generates, it is possible to discern a more lasting value in these articles. As is explained in the last article, the round table can be seen as a contribution to the formulation of a new philosophy of research in ecology and evolution for which the authors can justly be proud.

COMPETITION AND THEORY IN COMMUNITY ECOLOGY

JONATHAN ROUGHGARDEN

Department of Biological Sciences, Stanford University, Stanford, California 94305

Submitted September 8, 1981; Accepted February 3, 1982

This essay is about competition as an ecological and evolutionary force, and about theory in community ecology, particularly competition theory. It responds to Connell (1980), Strong et al. (1979), and Connor and Simberloff (1979) (who criticize the evidence for competition, and the theoretical attention that competition has received.) These criticisms imply that competition theory, including its extension to the coevolution of competitors, is irrelevant to natural processes and is unworthy of testing regardless of whether the testing is feasible. As someone active in community theory, I wish to explain why these criticisms are unfounded and what the nature and function of theory is in community ecology.

The essay is in three parts. The first is philosophical and concerns how one establishes an empirical fact in science. The essay begins with a philosophical discussion because the criticisms spring from what I feel are untenable positions in the philosophy of science. The second part details some serious technical flaws in the proposals of the critics. The third is an account of what theory in community ecology is, what it is trying to do, how it should be judged, and what some of the sources are of misunderstanding between theoretical and empirical points of view.

PHILOSOPHY

Some Initial Propositions

Consider some propositions about how facts are established in science. They provide a point of departure for evaluating philosophical aspects of the papers by Connell (1980), Strong et al. (1979), and Connor and Simberloff (1979).

We establish an empirical fact in science in the same way that we establish ordinary empirical facts during our everyday lives.

The way we establish an ordinary empirical fact is by building a convincing case for that fact.

In our everyday lives we rarely abide by formal rules to tell us how to build a convincing case; we use our native abilities, common sense, and experience in building a case for, and evaluating, claims of fact.

As scientists, we rarely abide by formal rules in establishing scientific facts. To establish a fact, we develop a case for that fact by appealing to the common sense and experience of people, most of whom are other scientists. Whether a case is convincing depends both on what is in the case, and on the knowledge, bias, and dispositions of the people to whom it is submitted. A finding is accepted as fact only as long as, and to the extent that, people remain convinced. Acceptance is neither complete nor eternal for a factual claim.

The truth of a factual claim is eternal; it either did or did not rain in Rome on the Ides of March in 100 B.C., and grassquits either did or did not eat the same types of food as anoles during the dry season of 1981 on St. Eustatius. In accepting or rejecting such claims, we evaluate the evidence offered on their behalf. Yet the degree to which we are convinced by the evidence depends on our experience at that time, and can change as our experience changes.

Any belief that scientists establish facts with more certainty than we can in our everyday lives is a delusion. Our distinctive activity as scientists is that we encounter and experience phenomena that are remote from our everyday lives, or that are overlooked during our everyday lives; but the way we try to understand these phenomena is with abilities whose credibility originates in everyday circumstances.

As scientists we use experimental setups, specialized equipment, and statistical techniques more often than we do in our daily lives; but all this is a matter of degree, not kind. In our daily lives we test the speed limit, sample clothes, alter recipes, and so forth; all activities with parallels in the practice of scientific inquiry.

I am not saying that the practice of scientific research is identical to what we do in our everyday lives. For example, as scientists we usually try to be more careful in our investigations than we might be about matters of casual and nonprofessional interest. Yet, the credibility of our reasoning and of our approach, as scientists, rests on their being referable to some analogous nontechnical situation in our everyday lives.

There can be a sense of uneasiness about the idea that scientists evaluate claims of fact with their common sense abilities rather than by applying some set of rules. It may seem that common sense is no more than uninformed prejudice, and that people come to conclusions about matters of fact in their everyday lives without being thorough, and by uncritically including evidence of dubious quality and relevance. But this is all part of the human condition; we cannot escape our humanity by passing laws against sloppiness and irrelevancy. Instead, we must remain free to evaluate such laws themselves against our experience and common sense.

This view of what scientists do is close to that of Huxley (1894) and also consonant with contemporary philosophers of science who emphasize the multiplicity of criteria used in the evaluation of scientific claims, as reviewed, for example, in Suppe (1977).

No Proof for Competition and Coevolution

In contrast with scientific findings, a proof that is valid according to formal rules eternally establishes a new theorem in logic and mathematics. Constructing the proof is a creative human activity, but acceptance is based on whether formal rules of validity have been correctly applied in the proof.

Connell (1980) presents a table of experiments that he insists be followed in order to determine whether the coevolution of competing species has led to niche separation. Connell (1980, p. 135) writes that the experiments are "it seems to me, both necessary and sufficient" to demonstrate that competition is present and that the niches of each species have a genetic basis. The meaning of this justification is unclear. The "it seems to me" might legitimately indicate what Connell would personally find convincing. The "both necessary and sufficient" misleadingly suggests, first, that there is an element of logical certainty compelling the acceptance of a conclusion based on Connell's protocol (sufficiency), and second, that no other protocol is satisfactory (necessity).

Following the formal rules of logic and mathematics actually guarantees both that a theorem is valid and that the theorem is accepted as valid.

In contrast, there is no table of experiments, no matter how well thought out, which, when followed, guarantees that competition actually occurs in a system and also that the finding will be accepted. At best, following a table of experiments would produce a case convincing to many people. Even here we reserve the right, for each situation, to make our own reasoned evaluation of the appropriateness of the protocol and of the clarity of the experimental results before accepting the conclusion.

A Litmus Test for Competition or Coevolution?

Nonetheless, perhaps following a general protocol could produce a case that is convincing to virtually all reasonable people. If we insert litmus paper into a solution, any solution, and it turns pink, then we conclude the solution is acidic. It is conceivable, however, that the paper is defective or that there is pink pigment in the solution to begin with. If so, then the conclusion that the solution is acidic may be incorrect. Yet without any evidence that something is amiss, it would be unreasonable to withhold agreement that the solution is acidic.

Could we devise a similarly general protocol to detect competition and coevolution, a protocol whose results are not necessary and sufficient, in a deductive sense, to guarantee the existence of these processes, but whose results are conclusive in a practical sense? I think not, but it is too early to tell. The circumstances seem too varied. The mechanism of competition varies as does the population structure that those mechanisms must affect.

No Logical Primacy to Null Hypotheses

Connor and Simberloff (1979) and Strong et al. (1979) propose procedural rules for the study of competition. They insist that scientists begin by defining many alternative hypotheses for the thesis they are investigating. For example, if one is investigating whether the coevolution of competitors has led to niche divergence, various alternative hypotheses are that predation somehow causes the divergence (e.g., the prey are evolving a size escape in opposite directions), that divergence results from a pleiotropic correlation with the evolution of other aspects of the phenotype, and so forth. Second, they insist that a special alternative hypothesis be singled out as the "null" hypothesis. The main feature of the "null" hypothesis is a proscription against population interactions. Third, they insist that the "null" hypothesis be falsified first (in time) before moving on to consider the alternative hypotheses that do include population interactions other than competition, finally ending up with a direct look at hypotheses involving competition itself.

There is certainly no objection to considering alternative hypotheses. The need to do this is obvious, and does not require a sophisticated philosophical justification. The objections are to a rigid sequencing of research activities according to rules, and to singling out a special alternative hypothesis as the "null" hypothesis.

Strong et al. (1979, p. 910) misuse philosophical jargon; they write, "we propose another possibility with *logical primacy* over other hypotheses . . . this is the null hypothesis that community characteristics are apparently random [my italics]." This is not logic; this is simply an unjustified assertion.

The "null" hypothesis of these critics is not simply the negation of the hypothesis being investigated. Clearly, establishing strong interspecific competition is equivalent to falsifying the absence of strong interspecific competition. If one calls the negation of a claim a "null" hypothesis, then, by this definition, to establish a claim one must falsify a "null" hypothesis. This is a trivial tautology. In contrast, the null hypothesis in Strong et al. (1979) is that there exist unspecified random processes in ecological communities that cause communities to be as they are.

In evaluating philosophical claims about how to do scientific research, it is helpful to return to everyday circumstances. Suppose you are sitting by a window in your living room with a guest. You can see your driveway through the window while your guest cannot. Your guest says, "I see the shadow of a car on the lawn, so your car is parked in the driveway." Obviously, this statement jumps to a conclusion that is not fully warranted by the data. Yet the conclusion may be true nonetheless, and the data that are cited may actually be enough to convince you. One alternative hypothesis is that you parked your car at the front curb. If you were not convinced, however, and wanted to verify the conclusion, you would first look directly at the driveway. You would not avert your eyes from the driveway in order to begin with an inspection of the front curb. Clearly, one is not logically compelled to start with an alternative hypothesis before getting around to the claim that is at issue.

Furthermore it is preposterous to single out one of the possible alternative hypotheses for special attention by calling it the "null" hypothesis. What is our null hypothesis here: that you have no car; that there is no car in the driveway; that there is no car on earth; that shadows flicker across the lawn by chance alone? There has to be an independent justification for the selection of a null hypothesis. Simply declaring some hypothesis to be logically prior is a presumption.

Moreover, logical priority has nothing to do with temporal priority. Logical priority does not, even when justified, imply that people should investigate the logically prior hypothesis before investigating other hypotheses. The temporal sequence of investigation is a practical matter, not a matter of logic.

The Popperian Alignment is Unjustified

Connor and Simberloff (1979) and Strong et al. (1979) are explicitly philosophical in proposing their rules. They claim to be descendants of the philosophy of Popper (1968). In brief, the philosophical idea is that one cannot directly confirm a scientific claim; instead, one concludes that a claim is true only after falsifying alternatives to the claim and failing to falsify the claim itself.

The critics do not present a clear and cogent justification for this philosophical alignment. Strong et al. (1979, p. 909) offer only the statement that modern science "has made much of its progress by attempting to disprove universal hypotheses like competition." What is not clear is whether Strong et al. (1979) find an inherent problem with experiments that directly detect competition, or whether they believe that progress will occur faster if scientists investigate alternative hypotheses before getting around to the principal thesis of interest. The latter position is philosophically innocent, and probably false. The former position is philosophically dubious.

More fundamentally, Connor, Simberloff, and Strong are attentive to their philosophical alignment during the course of their scientific research. I feel that this can lead both to bad science and to an abuse of philosophy.

Philosophy is an ongoing field; the perceptions of philosophers change as thought accumulates on research issues in philosophy. The important early research of Karl Popper is not necessarily among the best thinking currently available in the philosophy of science, a field that is just beginning to understand the differences among the various natural sciences (see, e.g., Beckner 1959; Hull 1974; Suppe 1977). It seems premature to adopt a philosophical alignment in this area of philosophy.

Some philosophy identifies and clarifies the main ideas in a subject; this is a "foundational" inquiry. It has little impact on the practice of the field being investigated. Research on the meaning of the infinitesimal does not affect the utility of the formula for the derivative of a polynomial in science and engineering,

$$d(x^n)/dx = nx^{(n-1)}.$$

Philosophical research is not irrelevant to this formula; it may lead to a different (and better) interpretation of its meaning, to a generalization of the concept of a derivative, and so forth. Yet it is a mistake in a scientific paper to fail to use this formula in a standard way because of a philosophical position on the meaning of the infinitesimal. Similarly, it is a mistake to act in ways contrary to good judgment when investigating a scientific claim because of an alignment with a position in the philosophy of science.

Popper's philosophy seems to be prescriptive; not about how scientists do work as much as about how they might work. Popper and others writing earlier in this century were interested in whether the certainty of scientific findings could approach those of mathematical findings. This may be interesting, although unnecessary. The potential danger, however, is that following Popper's dicta may lead to results that do not speak to the concerns of the actual scientists who have to be convinced.

Perhaps the most difficult issue facing both philosophers of science and interested scientists is to explore the role of the philosophy of science vis-à-vis science itself. Is the philosophy of science a metascience? By adopting alignments are scientists unwittingly participating in an experiment on how to do science? Is the best alignment to be gauged by the amount of scientific progress it produces? If so, the experimental design needs further thought.

If the philosophy of science is not a metascience, what is it? This question has been faced before in another guise. Here is Wittgenstein (1958), a giant in twentieth century philosophy, writing about the relation between formal logic and the natural language that people actually use for reasoning and for communicating with one another.

> In philosophy we often *compare* the use of words with games and calculi which have fixed rules, but cannot say that someone who is using the language *must* be playing such a game. —But if you say our languages only *approximate* to such calculi you are standing on the very brink of a misunderstanding . . . as if it took the logician to shew people at last what a proper sentence looked like.'' [Paragraph no. 81, p. 38, with original italics.]

> Philosophy may in no way interfere with the actual use of language; it can in the end only describe it. For it cannot give it any foundation either.
> It leaves everything as it is. [Paragraph no. 124, p. 49.]

Thus philosophy may also be descriptive; about how scientists do work, how they draw conclusions, what they count as evidence, whether the findings are self-consistent, whether its language contains terms that can be operationally defined, and so forth. We are the source material for this type of inquiry. We compromise ourselves when we abandon our natural talents in adhering to a philosophical alignment.

TECHNICAL CRITIQUE

Connell's Protocol

Connell's (1980) paper concerns whether coevolution between competing species is common and important. Connell wants the demonstration of coevolution between two competitors to consist of: (1) evidence of evolutionary divergence of the two competitors from the fossil record or other historical sources; (2) experimental evidence for present-day competition between the coevolved species based on observation of niche compression/expansion in a scheme of transplant/removal experiments; and (3) experimental evidence for a genetic basis to the species differences that are also based on observations of niche compression/expansion in a scheme of transplant/removal experiments. I do not think Connell's table of "necessary and sufficient" experiments is workable, convincing, or even on the right track.

A. If coevolution does occur between competing species, then divergence through time is only one possible outcome; indeed, a rather unlikely one. According to the theory being tested, whether this possibility is realized depends on the initial condition. Consider an island with one species, and visualize a resource axis of one dimension. In sufficient time the niche position of the species should shift to the point corresponding to the peak of the carrying-capacity function. Now consider an invader and suppose its initial niche position is to the right of the resident. If it is sufficiently far to the right, then both species will shift to the left. Then they may equilibrate, with the original resident located to the left of the peak position and the invader to the right of the peak position, although not as far to the right as the position at which it originally invaded. Only if the invader enters very close to the resident's niche position will there be a divergence such that the resident shifts to the left while the invader shifts to the right. A position close to an established resident, however, is an unlikely place for a successful invasion. In the more probable scenario, the fossil record would indicate a parallel evolution of both members. (Of course, the argument is the same with the directions reversed if the invader enters at the left of the resident.)

B. Connell's protocol focuses on observation of niche compression and expansion. The relationship of these observations to shifts in niche position is unspecified. Hence these observations are possibly irrelevant to the hypothesis being tested.

C. In the special case when competition does produce divergence over evolutionary time, the case that Connell is trying to find, the residual competition after divergence is low and hence should necessarily tax the resolution of his experimental system. Thus, Connell, himself, must detect the "ghost of competition past" if he is to follow his own protocol. Hence this recommendation is unworkable. An unworkable recommendation biases the protocol against the hypothesis being tested and might be taken to imply that the hypothesis itself is not testable; but it is the protocol that is flawed not the hypothesis.

D. Simple transplant/removal experiments hardly begin to assess whether there is a genetic basis for the difference between two populations of the same species that are sympatric and allopatric, respectively, with a presumed competitor. The important issue is whether there is variation that is heritable, i.e., capable of responding to directional selection (see Boag and Grant [1978] and Van Noordwijk et al. [1980] for examples of how the heritability of ecologically interesting traits in natural populations has been determined). Alternatively, a selection experiment could be conducted on the relevant traits to see if they do respond to selection. Indeed many selection experiments for size traits have been done over the last 50 yr with generally positive results.

The Context of Connell's Protocol

Connell's (1980) paper places too much faith in the conclusiveness of certain field experiments for studies of population dynamics. The general problem in interpreting experimental results from the intertidal systems that Connell cites is that interindividual interactions are detected, but the population-dynamic consequences of these interactions are unknown. Many sessile marine invertebrates have a population structure involving a long-lived, pelagic, larval phase. The range of the population may be very large. Data on the interactions of individuals at a few study sites within the species' range, and observed within a scale of a few square meters, have an uncertain connection to population and community processes. This connection needs to be investigated with other methods, including the use of biogeographic and distributional data.

Connell's recommended field experiments are biased toward the detection of interference mechanisms between individuals. It may seem that experimental studies of competition tend to establish the existence of interference mechanisms and fail to provide evidence of exploitative mechanisms. To the extent this is true, it is an artifact of studies focused on the observation of interindividual interactions. The existence of interference mechanisms does not rule out the presence of exploitative mechanisms as well. To the contrary, exploitative competition should cause the evolution of interference mechanisms. Without reference to some exploitative basis, it is difficult to explain why two species should actively interfere with each other at all.

Connell (1980, p. 136) maintains a view downgrading the importance of competition in nature that does not make sense. He believes that, under "benign" physical conditions, "natural enemies (predators, parasites, herbivores) tend to be more effective so they keep the populations below the level at which they compete." [Parentheses in original source.] It is not clear how this hypothesis could apply to all trophic levels at the same time.

Suppose there are two prey species, A and B, and a predator, P. Suppose that if P is removed, A causes the extinction of B. Next, with P present, suppose that P has no preference for either A or B; it eats both in proportion to their relative abundance. In this situation, B may still become extinct; it is because there is no differential effect of the predator on either prey species. Here is it fair to say that competition is unimportant? No, because it is the competition that may determine which species becomes extinct. Alternatively, suppose P has a preference for A, the superior competitor. In this situation B may coexist with A when P is present. Is it now fair to say that competition is unimportant? No, because B is influenced by its competitor, A, and this influence would be especially large should P be perturbed by pollution or some other factor. Thus, it does not make sense to downgrade competition as an important contributor to community structure solely because there are effective "natural enemies" in the system.

Connell (1980) concludes that the results of coevolution are likely to be restricted to systems of low dimension. (Here, the dimension of a system means the number of species in it.) I tend to agree, but for different reasons. Connell feels that in systems of high dimension, the species will not co-occur enough either in space or time for the potential interactions among them to yield evolutionary results. This may be true in some circumstances; no one knows. My studies of the theory for the coevolution of competing species show that even when species are assumed to be co-occurring, the coevolutionary process slows down as the dimensionality of the system increases (Roughgarden et al. 1983*a*; Rummel and Roughgarden 1983). There are two reasons. First, as more species are added to the system, the net selection pressure favoring any particular coevolutionary change diminishes. Consider several species on a resource axis. If one species moves, say to the left, it reduces competition with the species on its right, but adds to the competition with the species on its left, thereby reducing any net advantage to such a shift. Second, the species in a high-dimensional model affect one another through very long feedback loops, most of whose individual links are weak, causing the system to equilibrate only after a very long time. For example, consider again a species added to a place on a resource axis. It evolutionarily affects a species that is not adjacent to it primarily by influencing an adjacent species which, in turn, passes the effect on, though with diminished strength.

The tone of Connell's point is negative, as though a localization of the effects of coevolution to low-dimensional systems somehow negates the importance of coevolution as a process. To the contrary, although Connell dismisses the possibility, it remains extremely important if large-dimensional systems are decomposable into many coevolutionarily shaped low-dimensional subsystems within each of which strong interactions occur, but among which the interactions are weak or absent. It is not clear how large-dimensional systems can exist to begin with if they are not compartmentalized, because there is serious difficulty in achieving coexistence among a great many strongly interacting populations. The low-dimensional subsystems that probably exist within large-dimensional systems are excellent candidates for evidencing the results of coevolution. Furthermore, even if coevolution is slow in large-dimensional systems, some large-dimensional systems may exist in habitats that are permanent enough for the process to have attained noticeable results.

Connor, Simberloff, and Strong's Null Models

The basic error in the papers of Connor and Simberloff (1979) and Strong et al. (1979) is that the "null" hypotheses are empirically empty. No biological processes are exhibited that produce the distributions predicted by the null models. Hence, we do not learn anything by "falsifying" these hypotheses. These null hypotheses are irrelevant to whether competition influences natural communities.

The null models are stated as though based on sampling theory. Actually, the data analyzed are not a sample drawn from a population of data in the usual sense. Typically one would regard the data taken on an island as a sample of what was actually on the island. Then confidence limits would be applied to these data based on statistical sampling theory. Instead, island data, assumed error free, are regarded as a sample of some faunal source pool. Islands do not reach into urns and draw out their species. There are real processes that bring species to islands. It is these processes that must be modeled to determine what distributions of data are expected in the absence of competitive interactions among species.

Fabricating random communities by sampling with replacement from some source fauna does not bear the relation to ecology that the neutrality hypothesis does to population genetics. The neutrality hypothesis is about real processes. Genetic drift is a real process whose workings have been demonstrated in the laboratory and in the field. Similarly, natural selection is known to be a real process. What is being claimed by the neutrality hypothesis is that selection is largely a "purifying" selection as distinct from various schemes of "balancing" selection. The neutrality hypothesis in genetics has empirical content.

I stress there is no disagreement about the usefulness of considering alternative hypotheses. The point is that null models fabricated by rearranging species lists, and indices of species morphology, are not viable alternative hypotheses. Indeed, the biological stochastic processes of dispersal combined with population extinction do offer viable hypotheses for the formation of island faunas and these hypotheses are alternatives to those postulating a key role for population interactions during faunal buildup. But null models are not about processes at all, and so remain irrelevant to the study of how island faunas are formed, regardless of how well or poorly they compare with data. Moreover, following upon the philosophical discussion earlier in this essay, hypotheses based on stochastic processes of dispersal and population extinction share the same logical status as other hypotheses, and no methodological rule requires that stochastic models be investigated before any of the other hypotheses.

Null models misrepresent the concept of a null hypothesis in statistics. A null hypothesis in a statistical test is a model, that is, a simplified picture, of how data are taken and what they are taken from. In developing a statistical test, one calculates what data are anticipated from this model. If the actual data are sufficiently different from what is anticipated, then the null hypothesis is rejected. Statisticians justify the null hypothesis in a statistical test by referring to the actual sampling practices of people who gather data, and by developing theorems, like the central limit theorem, that describe the characteristics of the populations being sampled (often a population of sample statistics). In fact, different formulas are available for placing confidence limits on population size estimates from mark-recapture data based on different pictures of the sampling process that is occurring when people conduct mark-and-recapture work under field conditions. A null hypothesis in statistics is a justified model of a sampling procedure. It is not a hypothesis that the world has no structure.

Strong et al. (1979) imply that one must investigate pattern before one is licensed to pursue research into the processes that may have caused the pattern. This is a fallacy. Sometimes it is obvious that a process is occurring. Knowledge of that process may aid in discovering its consequences.

Grant and Abbott (1980), Hendrickson (1981), and Diamond and Gilpin (1982) have shown that the null models lack statistical power (inability to distinguish the "checkerboard" pattern) and suffer methodological irregularities (the "dilution effect" and the incorporation of competition-dependent information). The inclusion of competition-dependent information in a noninteractionist model is not necessarily fatal, however, provided the degrees of freedom for the model are correspondingly reduced.

Strong et al. (1979) confuse the character displacement possibly resulting from the coevolution of competitors with the niche separation resulting from selective invasion at niche positions where competition is relatively low. These processes are quite different. The morphological data are taken solely from continental birds, and yet are applied to species lists from the Tres Marias islands. Obviously, these data cannot speak to whether species have changed after arriving on the islands. These data may not even be relevant to selective invasion if some of the early invading species changed substantially before the more recent invasions.

The Conservative Posture

The papers of Connell (1980), Connor and Simberloff (1979), and Strong et al. (1979) are biased against the existence of competition. They offer no justification for this bias. In any particular field system, there are two possible errors of assertion. First, one may assert that there is competition when in fact there is not (a type I error), and second, one may assert that there is no competition when in fact there is (a type II error). The critics are severely biased in favor of committing a type II error in order to avoid a type I error. To justify this bias, the critics should exhibit that a far greater harm results if we accept the presence of competition when it is really absent, than if we reject competition when it is really present. Still another approach would be to withhold judgment until the case is more complete.

The Case for the Coevolution of Competitors

How, then, are we to determine if the coevolution of competitors occurs? We see if we can build a convincing case. If we cannot build a convincing case in some system or a close analogue then we should dismiss the matter in that system. Better yet, if competition is unimportant it would be more informative to find out what is. A convincing case should include on-site experiments, together with biogeographic and distributional data, and data addressing viable alternative hypotheses. An analysis of morphological data and species lists may suggest the presence of competition. It is better, though, to regard morphological data and species lists only as a description of what is to be explained rather than as evidence of any particular process.

The mark of a convincing case is that the proposition of interest has been examined fairly from every angle. The case must be thorough, and yet balanced, with no particular angle pushed to the point of overkill.

The case should be evaluated on the basis of its content and not on the basis of philosophical and methodological doctrine.

THEORY

Something of a tone of righteous indignation in Strong et al. (1979) and of ridicule in Connell (1980) seems to tinge the antitheoretical rhetoric. Strong et al. (1979) assert that textbooks have uncritically accepted an interpretation of community structure based on competition theory as a "paradigm" (p. 909) and that the "enthusiasm" (p. 897) for this paradigm is so strong that alternative explanations and contradictory evidence are ignored. They claim there is a "lack of reciprocity between material and theoretical effects" in the study of competition (p. 909). And Connell (1980, p. 132) writes, "Despite all the theoretical attention it [the coevolution of competitors] has received, there remains a real question as to how much this notion applies to real communities." Has competition theory really escaped from biological control?

To begin, many authors of ecology texts in North America, Europe, and Australia hardly embrace competition theory. Most textbook authors view competition theory with a wary eye, if at all. Furthermore, the frequency and skill with which field experiments are done in community ecology is continually increasing and important alternative hypotheses are being addressed in such studies. It would be a clear disservice to students in ecology today if textbooks failed to mention: (1) the phenomenon of resource partitioning; (2) the "checkerboard" distribution pattern of species of the same body size; (3) the regularity of the degree of difference in body size among the species who do coexist; and (4) that a possible explanation for these facts is provided by competition theory.

Perhaps more to the point, what is the purpose of theory?

Characteristics of Theory

The purpose of most models in community ecology is to simplify.

Making a simplifying model typically involves three steps. First, we agree to consider a subset of what is known about a system. Second, we fill the gaps in the simplified description with assumptions that may actually be false if taken literally, but which are hoped somehow not to be too misleading. Also, even the simplified description may contain points that are too complicated to deal with and so are replaced by other statements that are more tractable. Third, we derive predictions. Sometimes the derivation serves only to show that some features of the simplified description follow from others. In this context the derivation reveals interrelationships among the features of the system. Other times the derivation leads to a new proposition that no one has considered before. The simplified description of the system, together with the predictions, comprise a model. A set of models sharing many assumptions in common is a theory.

Why would someone want to develop a simplifying model? Because experience shows this is useful. In school we learned the concept of mechanical advantage from the study of simple machines like the frictionless pulley. Anyone who has hoisted the engine from a car knows that as the number of pulleys increases, the length of rope that must be pulled also increases, but the strength needed to pull the rope decreases. Moreover, when the pulley is well oiled, the model of a frictionless pulley is actually quite accurate. Similarly, network theory, the most important theoretical tool in electrical engineering, is based on simplified models of electrical components. Experience shows that, although simplifying models incorporate dubious assumptions, like that of no friction, they are useful as aids to understanding what is going on and often serve as guides to the quantitative description of systems.

A simplifying model is vulnerable. It differs from a summarizing model whose purpose is to include everything that is known about a system and that is continually updated as more is learned. One cannot falsify a model that is continuously updated; it has, so to speak, no degrees of freedom.

The predictions of a model are robust if, as a historical accident, they happen to have been derived from premises that later research shows were unnecessarily restrictive.

In fact, most theoretical results are derived for the first time in models that have been unnecessarily simplified. How do we know to what extent results are robust? This is very difficult. The mathematical approach is, first, to try to generalize the original derivation somewhat and, second, to bound the robustness by developing mathematical counterexamples. The empirical approach is, to me, more interesting and it consists of deliberately looking for the prediction of a model in a situation that is known to be somewhat different from what is described by the assumptions of the model. If the predictions check out, then we have a clue that they may be more robust than originally believed.

The results of a model are structurally stable if they change quantitatively in proportion to quantitative alterations to the premises. The idea is that a small change in the premises produces a correspondingly small change in the results. Some technical criteria are available for ascertaining structural stability. For example, it is known that if the dimensionality of a dynamical model is high, then the global flow of the trajectories is not likely to be structurally stable; but the flow of trajectories in the neighborhood of a locally stable equilibrium usually is structurally stable.

Theory of Interspecific Competition

Niche theory offers a simplified picture of how competition can produce and maintain a pattern of resource partitioning among coexisting competing species. It also addresses the phenomenon of zonation, including the length of the overlap zone near the borders of species that replace one another along geographical gradients. Its results for low-dimensional systems are structurally stable and moderately robust from a mathematical standpoint. At present, niche theory offers very few predictions concerning large-dimensional systems (see review in Roughgarden 1979).

The research of Diamond (1975), so strongly criticized by Connor and Simberloff (1979), remains a very important exploration of the possible consequences of competition in high-dimensional systems. His study, and others, do not directly test or support competition theory as much as offer suggestions of what to look for as possible predictions from high-dimensional theory when it is developed. Such suggestions are essential if theory is to present results of biological relevance. Most of Diamond's suggestions about how competition causes patchiness have been sustained by an analysis of the competition equations themselves (Roughgarden 1978). Other suggestions, including the possible existence of assembly rules and forbidden combinations of species are clearly relevant to the multiple, simultaneously stable, boundary equilibria that are possible in high-dimensional systems with strong competition.

Testing Theory

Summarizing models are tested by direct comparison with the data being predicted. If the model parameters are estimated from, say, the data of the last 5 yr, then the model is used to predict the new year's data return. If the new data are impressively different from the predictions, then the model itself is reworked, or at least the parameter estimates are redone. Meanwhile the outdated model is automatically discarded. Testing a model in this sense is called validating a model.

For simplifying models the matter is more subtle. A simplifying model must be somewhat unrealistic to begin with; simplification is what it is all about. It must remain somewhat unrealistic if it is to continue being a simplification rather than a summary. Arriving at an acceptable degree of simplification involves human judgment.

Testing a simplifying model means establishing whether the picture supplied by the model is, by and large, a correct account of what is going on in a system.

There are three reasons why a simplifying model may fail to provide a correct account of what is going on in some system. (1) The model may be about processes that do not occur in the system. (2) It may be about processes that do occur, but that are relatively unimportant when compared to other processes that have been omitted from the model. (3) The model may be about processes that do occur, but may provide a severely distorted characterization of those processes.

A model is biologically general if the picture it presents is correct for many systems. Biological generality rests on either a universal mechanism (as in Mendelian inheritance) or on the existence of a large variety of mechanisms that work in the same way.

The fate of a simplifying model is determined by people's interest. Does a model help in planning research or in understanding results, or even in understanding other models? If not, the model does not hold the interest of active scientists and falls into disuse. A similar fate awaits empirical findings that, although correct, turn out to be inconsequential.

It is a distraction to focus on the model rather than on processes in an actual system. For example, on the island of St. Maarten there is strong present-day competition between two species of lizards, *Anolis gingivinus* and *A. wattsi*. I hypothesize that the absence of *A. wattsi* from sea-level habitat on St. Maarten represents competitive exclusion by *A. gingivinus*. The Lotka-Volterra competition equations help us to visualize how competitive exclusion occurs as a population process, and thereby aids in planning the research to determine if this hypothesis is true. However, what we are testing is not so much the Lotka-Volterra model, but whether there really is present-day competition strong enough to cause competitive exclusion on St. Maarten. The empirical issue, and not the model, is what is significant. If indeed there is competitive exclusion on St. Maarten, then we will have a key point to the discovery that the coevolution of competitors can lead to extinction. Such a discovery would show that the coevolution of competitors may not necessarily lead to character displacement with an accompanying reduction of competition, as is widely believed. Furthermore, it was the mathematical theory for coevolution that sensitized us to this possibility.

Theory does not substitute for the knowledge of actual systems any more than architectural drawings substitute for a house. Yet we cannot build a house without architectural drawings.

Nonetheless, today we do plan and carry out almost all field research without aid of the insights that simplifying models provide. This, however, may be a transient state in the history of ecology. With more experience in using theory we may come to ask questions about processes that now seem hopelessly complicated and may come to devise more informative experimental designs than we presently envision.

Living with Theory

There is antagonism among many ecologists toward theory, and some of it arises, I suspect, from the fear that ecological theory is considered the "foundation" of ecology. Some sciences, like physics, are hierarchical, and physicists speak of theoretical axioms, laws, and of "truths" that have been derived from such theory. In a hierarchical field, it is conceivable that a misdirected theory could divert the entire field away from a commonsense evaluation of its own empirical findings; if so, this is a legitimate fear. Ecology does not have such a hierarchy now, I doubt it ever will, and hope it never does. It is difficult to imagine what could ever qualify as a "law" in ecology. Ecological theory is no more than a collection of tools. A useless model should be discarded like a broken chisel.

Another source of antagonism toward theory springs from a conflict between the ethic of perceptiveness among naturalists and the need to simplify among theoreticians. Among naturalists, failure to note detail is, in the extreme, equated with a lack of perceptiveness. In this context, simplification becomes synonymized with sloppiness, and with impatience, as of a novice who jumps to conclusions before a phenomenon has been thoroughly observed. On the other hand, the steadfast refusal to simplify ultimately leads to a glorification of the particular, to a celebration of the personality of each species and their interactions in each habitat.

There is a craft to developing theory. Solving for equilibria, for example, tells one about the mathematical structure of a model. In solving for equilibria, one is not saying the world is constant. To the contrary, one way to model environmental stochasticity is to introduce stochastic variation into the parameters of a model that was originally solved using parameter constants. Similarly, it is good practice to explore the premises of a model one by one and not to jump in with all of them at once. Doing so gives the appearance of leaving out important factors. The early analysis of a model is often sufficiently interesting to warrant publication in a theoretical journal, thereby giving the impression of completion. Still more discord sounds when someone's cherished fact is not incorporated into a simplifying model.

Theoretical research has an integrity of its own; it does not march in lock-step with empirical research. Nonetheless, theoretical ecology is ultimately valued by its utility in understanding ecological processes in nature.

There is a problem, though, that is more than a misunderstanding of the nature of the craft. It is that, for a theoretician, the natural object of study usually corresponds to only one of many hypotheses for a phenomenon. Suppose one is interested in what causes the observed body sizes of male lizards. There are several hypotheses to consider, involving resource use, sexual selection, mechanics of body growth, to name a few. Each hypothesis may require separate theoretical development, and they may not be equally tractable. A theoretician may think his job is done when one of these is satisfactorily modeled, that to take up another hypothesis is to take on another job. From an empirical standpoint the job is to build a fair case about the phenomenon itself. And if the hypotheses have received unequal theoretical treatment it is hard to develop a case that is fair to all reasonable hypotheses. The same problem exists when the hypotheses differ in their tractability for testing in the field. In short, it is hard to build a fair case when some hypotheses are easier to study, or are more studied, than others.

If there is any villain on the scene, it is the set of popular impressions about "what theory says". These impressions are always naive and usually incorrect concerning what theory actually does say.

The place of ecology within the sciences depends on the perception by scientists from other fields and by the public that progress is being achieved in ecology. This perception becomes the basis for science policy decisions concerning faculty billets and research support. The advance of theoretical ecology during the 1970s is an important event in contemporary science; it has begun to attract serious attention from other areas of science and from mathematics; and it has contributed to enhancing the stature of ecology as a result. This growth of theory, together with the increased use and understanding of experimental methods under field conditions, and the determination of biogeographic patterns to community structure, are evidence of substantial progress in community ecology during the last decade.

Theory in the 1980s

Where does community theory go from here? The theory of the 1970s explored models formulated in the 1960s and earlier. The theoretical content of these models and their relationships to one another and to models in population genetics were largely unknown at that time. During the 1980s the challenge is to formulate new models that can serve as useful simplifications for community processes in systems to which current theory is largely irrelevant. Hickman (1979) has eloquently called for such theory in annual plant communities. The rocky intertidal and herbivorous insect communities merit similar attention.

Another challenge is to extend local community theory to take account of migration-coupling among habitats. The number of species at a local site, especially on continents, often cannot be explained by observations taken within that site. Instead, the local species diversity reflects an interaction between the regional habitat diversity and the degree of habitat segregation practiced by the species. The habitat segregation that evolves probably depends on the geometry of the habitats available in the region, and on the long-term temporal stability of those habitats.

Still other important topics include the specialization of coevolutionary theory for the systems in which coevolutionary phenomena are readily studied, the incorporation of size and age structure into models of density-dependent population dynamics, the development of mechanism-based submodels for the parameters in population models and for the constraints used in optimization models, and the exploration of how population interactions can theoretically control ecosystem processes.

CONCLUSION

This essay was provoked by the insistent claims of Connell (1980), Connor and Simberloff (1979), and Strong et al. (1979) that competition and the coevolution of competitors are not real and important processes in nature, and hence, that theory for these processes is not worthy of testing. These authors are wrong. Both competition and coevolution among competitors occur, and are extremely important in some systems.

The genus, *Anolis,* contains 5% to 10% of the lizard species in the world today, and about half of all anole species reside in the West Indies. Anoles are very common; 0.5 to 2 individuals per m^2 is typical. They substitute for ground-feeding insectivorous birds, and, as major consumers, probably determine the abundance of many arthropod species, and probably set the flow rates for matter and energy through their terrestrial ecosystem.

Collaboration with S. Pacala, J. Rummel, and others has established experimentally that there is strong present-day competition between the two *Anolis* lizard populations on St. Maarten (see Heckel and Roughgarden 1979; Roughgarden et al. 1981; Pacala and Roughgarden 1982*b*; Roughgarden et al. 1983*a*, 1983*b*; Pacala et al. 1983; Pacala and Roughgarden 1983; Adolph and Roughgarden 1983; and Roughgarden et al. 1984). Excellent indirect evidence existed by 1977, but now both the effect of *A. gingivinus* against *A. wattsi* and the reciprocal interaction have been demonstrated with repeated experiments. Studies of physiological ecology and microclimate ruled out an alternative hypothesis involving thermal requirements. From a comparison with St. Eustatius, the magnitude of the competition has been experimentally shown to vary with the difference in body size (the more similar the body sizes, the stronger the competition). Also, the possible importance of predation and competition from birds and other taxa has been studied. The pair of species on each island bank with two species has been together at least since the Pleistocene, and all the anole populations in the Lesser Antilles have evolved taxonomic endemism. It is now practically certain that anoles in the Lesser Antilles have coevolved as competitors with one another.

Mathematical theory for the coevolution of competing species provides a picture of the process of faunal buildup on islands. This theory seems to account for the regularities and also for all exceptions in the pattern of species diversity and body sizes for anoles throughout the eastern Caribbean (Pacala and Roughgarden 1982*a*; Roughgarden et al. 1983*a*; Rummel and Roughgarden 1983). This theory is about real processes. It provides a genuinely useful simplified picture of what may be happening in the *Anolis* system of the eastern Caribbean.

ACKNOWLEDGMENTS

I thank many people for their interest and encouragement in writing this essay, and for their helpful comments on the manuscript. I thank especially Charles

Baxter, Michael Bratman, Paul Ehrlich, Anne Harrington, Richard Holm, Kent Holsinger, Harold Mooney, Stephen Pacala, John Rummel, John Thomas, and James Watanabe. I also thank Deborah Rabinowitz for showing me the reference to Huxley (1894).

POSTSCRIPT

The original draft of the essay above is now over two years old. Circumstances have changed for the better since it was written. Joe Connell's contribution reveals that the existence of competition, and, to a lesser extent, the coevolution of competitors is accepted for some natural communities. I agree with Don Strong that competition seems unlikely to be important in the many arthropod systems he cites, and repeat my original call for the development of models that offer useful simplifications for systems to which existing community theory is irrelevant. Concerning competition theory, I can vouch for its utility in at least one system, and there is literally a world of difference between a theory that applies to one system and a theory that applies to no system. Moreover, George Salt's remarks testify to an increasing appreciation of theory as a tool to aid in ecological research; its track record, even in community ecology, perhaps the most difficult area in which to test theory, is slowly but steadily improving. I still part company with Dan Simberloff both on philosophical issues and on whether his null models can be counted as viable alternative hypotheses. Our differences notwithstanding, it is clear that we now know much more about ecological communities than we did a decade ago, and ecological theory has contributed to this progress. (J. R., May 6, 1983).

LITERATURE CITED

Adolph, S. C., and J. Roughgarden. 1983. Foraging by passerine birds and *Anolis* lizards on St. Eustatius (Neth. Antilles): Implications for interclass competition and predation. Oecologia 56:313–317.

Beckner, M. 1959. The biological way of thought. Columbia University Press, New York.

Boag, P. T., and P. R. Grant. 1978. Heritability of external morphology in Darwin's finches. Nature 274:793–794.

Connell, J. 1980. Diversity and the coevolution of competitors, or the ghost of competition past. Oikos 35:131–138.

Connor, E., and D. Simberloff. 1979. The assembly of species communities; chance or competition? Ecology 60:1132–1140.

Diamond, J. 1975. Assembly of species communities. Pages 342–444 *in* M. Cody and J. Diamond, eds. Ecology and evolution of communities. Harvard University Press, Cambridge, Mass.

Diamond, J., and M. Gilpin. 1982. Examination of the "null" model of Connor and Simberloff for species co-occurrences on islands. Oecologia 52:64–74.

Grant, P., and I. Abbott. 1980. Interspecific competition, island biogeography and null hypotheses. Evolution 34:332–341.

Heckel, D., and J. Roughgarden. 1979. A technique for estimating the size of lizard populations. Ecology 60:966–975.

Hendrickson, J. A., Jr. 1981. Community-wide character displacement reexamined. Evolution 35:794–810.

Hickman, J. 1979. The basic biology of plant numbers. Pages 232–263 *in* O. Solbrig, S. Jain, G. Johnson, and P. Raven, eds. Topics in plant population biology, Columbia University Press, New York.

Hull, D. 1974. Philosophy of biological science. Prentice Hall, Englewood Cliffs, N.J.
Huxley, T. H. 1894. Pages 361–375 in Darwiniana. Macmillan, London.
Pacala, S., and J. Roughgarden. 1982a. The evolution of resource partitioning in a multidimensional resource space. Theor. Popul. Biol. 22:127–145.
———. 1982b. Resource partitioning and interspecific competition in two two-species insular *Anolis* lizard communities. Science 217:444–446.
———. 1983. Population experiments with the *Anolis* lizards of St. Maarten and St. Eustatius (Neth. Antilles). Ecology (in press).
Pacala, S., J. Rummel, and J. Roughgarden. 1983. A technique for enclosing *Anolis* lizard populations under field conditions. J. Herpetol. 17:94–97.
Popper, K. 1968. The logic of scientific discovery. Hutchinson, London.
Roughgarden, J. 1978. Influence of competition on patchiness in a random environment. Theor. Popul. Biol. 14:185–203.
———. 1979. Theory of population genetics and evolutionary ecology: an introduction. Macmillan, New York.
Roughgarden, J., W. Porter, and D. Heckel. 1981. Resource partitioning of space and its relationship to body temperature in *Anolis* lizard populations. Oecologia 50:256–264.
Roughgarden, J., D. Heckel, and E. Fuentes. 1983a. Coevolutionary theory and the biogeography and community structure of *Anolis*. Pages 371–410 in R. Huey, E. Pianka, and T. Schoener, eds. Lizard ecology. Harvard University Press, Cambridge, Mass.
Roughgarden, J., J. Rummel, and S. Pacala. 1983b. Experimental evidence of strong present-day competition between the *Anolis* populations of the Anguilla Bank: a preliminary report. Pages 499–506 in A. Rhodin and K. Miyata, eds. Advances in herpetology and evolutionary biology—Essays in honor of Ernest Williams, Cambridge, Massachusetts. Bull. Mus. Comp. Zool. Harv. Univ.
Roughgarden, J., S. Pacala, and J. Rummel. 1984. Strong present-day competition between the *Anolis* lizard populations of St. Maarten (Neth. Antilles)'. In B. Shorrocks, ed. Evolutionary ecology. Blackwell, Oxford (in press).
Rummel, J., and J. Roughgarden. 1983. Some differences between invasion-structured and coevolution-structured competitive communities—a preliminary theoretical analysis. Oikos (in press).
Strong, D., Jr., L. Szyska, and D. Simberloff. 1979. Tests of community-wide character displacement against null hypotheses. Evolution 33:897–913.
Suppe, F. 1977. The structure of scientific theories. University of Illinois Press, Chicago.
Van Noordwijk, A. J., J. H. Van Balen, and W. Scharloo. 1980. Heritability of ecologically important traits in the great tit. Ardea 68:193–203.
Wittgenstein, L. 1958. Philosophical investigations (trans. by G. Anscombe.) Blackwell, Oxford.

ON HYPOTHESIS TESTING IN ECOLOGY AND EVOLUTION

James F. Quinn and Arthur E. Dunham

Division of Environmental Studies and Department of Zoology, University of California, Davis, California 95616; Department of Biology, University of Pennsylvania, Philadelphia, Pennsylvania 19104

Submitted March 23, 1982; Accepted April 27, 1982

The use of explicit hypothesis testing in ecological and evolutionary studies has been the subject of much recent discussion. In the past several years, a number of workers have been severely criticized for a failure to pose and test alternative explanations for patterns observed in nature (e.g., Abbott et al. 1977; Connor and Simberloff 1978; Grant and Abbott 1980; Simberloff 1978; Strong et al. 1979; Simberloff and Connor 1981) and several philosophical overviews have recently been published (Peters 1976; Levins and Lewontin 1980; Simberloff 1980; Strong 1980). The intellectual basis for this discussion is not new. In the early seventeenth century, Francis Bacon (particularly the *Novum Organum*, 1620) discussed the role of proposing alternative explanations and conducting explicit tests to distinguish between them as the most direct route to scientific understanding (Eiseley 1973). Popper had outlined his formulation of proper scientific method by 1934. In an influential paper, Platt (1964) characterized explicit formal hypothesis testing in science as "strong inference" and argued that it is a hallmark of virtually all scientific progress:

Strong inference consists of applying the following steps to every problem in science, formally, explicitly, and regularly:

1) Devising alternative hypotheses;

2) Devising a crucial experiment (or several of them), with alternative possible outcomes, each of which will, as nearly as possible, exclude one or more of the hypotheses;

3) Carrying out the experiment so as to get a clean result;

1') Recycling the procedure, making subhypotheses or sequential hypotheses to define the possibilities that remain; and so on (p. 347).

... For exploring the unknown, there is no faster method; this is the minimum sequence of steps (p. 347).

It seems to me that the method of most rapid progress in ... complex areas, the most effective way to use our brains, is going to be to set down explicitly at each step just what the question is, and what all of the alternatives are, and then to set up some crucial experiments to try to disprove some (p. 352).

In this formalism, correct explanations cannot be proven deductively, except by eliminating all possible alternatives, but incorrect explanations may be disproven by contradictory experiments or observations (Popper 1959, 1972, 1983*a*).

Popper (1959, but see 1983a, 1983b) argued explicitly that estimates of probability are unfalsifiable, and thus not subject to scientific test under criteria of disproof. In general, propositions not subject to rejection by contrary observations are denied status as scientific theories in the Popperian model. Useful empirical generalizations with known exceptions, including most biological "laws" seem to fall in this category. Perhaps the best known consequence of this viewpoint is Popper's celebrated denial that Darwin's theory of evolution by natural selection is a valid scientific theory (1972, 1976; see also Peters 1976). Clearly this argument in its simplest form poses major problems for ecologists and evolutionary biologists. Postulated ecological causes or relationships can rarely be strictly disproven, although they may often be shown to be unimportant or improbable. Tests of historical causality and experiments with evolutionary constraints are often impossible in principle to perform.

In practice, the logic of ecological and evolutionary research differs from the Popperian model in being largely inductive. Biologists usually rely on a methodology much akin to statistical hypothesis testing, in which potential causal processes are identified, and their probable contributions are evaluated, weighted, and tentatively generalized to other situations. Generally, the object of investigation is the proportion of observed variation that may be explained through the use of one or more predictors (e.g., food limitation, predators, soil nitrogen, etc.), and the proportion that is to be ascribed to "chance." The predictors of "statistical hypotheses" are chosen in part for reasons of simplicity, measurability, and tractability, and represent deliberately oversimplified caricatures of the assumed underlying processes. "Chance" refers less to true physical randomness than to the contributions of the large number of deterministic effects not included in the model. It is insufficiently appreciated that hypothesis testing in this sense is an inductive or even descriptive procedure, and does not correspond especially closely to the deductive logic of "strong inference." The use of the words "hypothesis testing" for both certainly has helped obscure the differences. The logic of statistical inference has been developed at length by Hacking (1965) and many others, and will not be reviewed here.

From the formal deductive model, Platt (1964) and others derive a prescription for proper and efficient scientific methodology: that potential explanations should be explicitly listed and incorrect ones systematically eliminated, leaving an ever dwindling number of possibilities within which the truth presumably must lie. The simplicity of this prescription and its apparent success in some physical sciences and experimental areas in biology belie the difficulty of its application to complex systems of multiple causality, such as those usually studied in ecology and evolution. Implicit in "strong inference" is an assumption that the competing hypotheses to explain observed phenomena are general, mutually exclusive, and, to some extent, exhaustive. Falsification by observation or experiment occurs only to the extent that a hypothesized cause of an observed phenomenon can be shown not to operate at all, and is informative only to the extent that the disproof may be generalized to other situations. However if many causes contribute to an observed pattern, none will be eliminated from consideration by a properly designed experiment. Generally, no single cause can be shown to account for all of the observed variation in patterns and processes in natural communities. The objective of investigation in cases of this sort is not to determine the single cause of a pattern, as no such cause exists, but rather to assign relative importances to the contributions of, and interactions between, a number of processes, all known or reasonably suspected of operating to some degree.

A number of epistemological and historical critiques of the usefulness of a strong inference model of scientific method now exist. The limitations of purely deductive logic and the doctrine of disproof of specific hypotheses as a prerequisite to progress have been developed at length by Lakatos (1970, 1974) and others. Some philosophers dispute both the possibility of objective criteria for scientific truth in many kinds of investigation (Feyerabend 1975) and the desirability of complete objectivity on the part of individual scientists (Campbell 1979). It is certainly true that the history and sociology of actual scientific advances often correspond poorly to the process envisioned in the hypothetico-deductive model (Kuhn 1970; Brush 1974). Nevertheless, many ecologists and evolutionists appear to accept "strong inference" as the proper model for investigation of complex biological phenomena.

In the ensuing discussion, we will be less concerned with formal method than with examples of specific applications to ecological research. We accept that formal methodologies, even if imperfect, are often useful in clarifying logical stuctures of arguments and in identifying avenues of thought that might otherwise be missed. However we believe that attempts to force study of highly overlapping mechanisms of ecological and evolutionary change into a rigid hypothetico-deductive mold have the potential to detract from understanding. We see three major classes of problems.

1. Formal hypotheses generally cannot usefully be posed in a way that allows meaningful disproof of a finite number of discrete possibilities. Possible contributing causes are not "hypotheses" of hypothetico-deductive reasoning, because in patterns with multiple causes, it is not possible in principle to perform "critical tests" to distinguish between the "truth" of processes occurring simultaneously.

2. Treating possible contributing causes as distinguishable hypotheses leads to univariate critical tests. However, the behavior of a multivariate process may not be safely inferred from any combination of univariate tests if there are strong interactions among contributing causes.

3. In the hypothetico-deductive formalism, understanding is only increased when a hypothesis is rejected. Thus it may be presumed that acceptance of ecological or evolutionary causality only conveys information when a converse "null hypothesis" of nonexistence of the cause has been rejected (e.g., Strong 1980); however propositions about causality in natural communities rarely have single or simply stated converses. In practice, reliable null hypotheses may often be impossible to construct, as we generally cannot deduce the nature of the expected patterns that would evolve in the absence of any given biological process. In most cases, even if a "null hypothesis" can be posed, it has no probability of being strictly correct, and a sufficiently sensitive critical test will necessarily lead to rejection. Thus testing a "null hypothesis" would appear to have no value in formal deductive logic. In practice, null hypotheses represent

reference points for measurement (an inductive procedure) rather than constructs of deductive logic. In this respect, they are similar to other biological models, such as optimality or Leslie matrix formulations, which also may provide useful approximations to some natural processes, but have no probability of being strictly true. The usefulness of any model in this regard depends upon the reliability of its formulation.

The first point seems self-evident. A moment's reflection will reveal that testable statements about causal relationships in ecology and evolution are virtually never posed in a way that makes "alternative hypotheses" mutually exclusive. This is particularly true when processes may vary over a continuum (e.g., the relative unpalatability of a mimic in Batesian vs. Müllerian mimicry). It may often be useful to classify continuously varying phenomena into a finite number of discrete categories. It does not seem reasonable, in a quest for methodological purity, to require the use of predefined categories in order to permit falsification of discrete statements. Attempts to force observations into categories erected for other purposes or circumstances can easily lead to overlooking unique characteristics of a particular situation. As an illustration of this problem, consider the proposition that competitive processes structure natural communities. The hypothesis that competition is the exclusive determinant of species' ranges and abundances can of course be rejected a priori, as can the "null hypothesis" that competition has exactly no effect. In practice, we hope to measure the relative impacts of competition and perhaps predation or soil structure, but it is not at all clear what kind of discrete hypothesis such an endeavor could reject. (We may, of course, conclude that the contribution of competition is too small to measure, but that is a matter of probable strength, not existence, of competition as a possible contributor to the observed pattern.)

The second point is closely related to the first. Effects of simultaneous processes often do not combine additively. Thus attempts to perform a series of univariate "tests" of individual causal factors may misestimate their actual contributions. Yet under a formal hypothetico-deductive scheme, interactions between mutually exclusive causes are not possible. Here again, we argue that appropriate biological methodology is often more analogous to standard statistical hypothesis testing than to "strong inference." Testing for nonadditive interactions is of course a standard part of multivariate experimental design and statistical analysis.

The third class of problems is related specifically to the use of "null hypotheses" in the study of natural communities. A number of workers have proposed the routine application of noninteractive "random" models as alternatives to models involving interspecific interactions (Caswell 1976; Connor and Simberloff 1978; Lawlor 1980; Strong et al. 1979; Simberloff 1978). Plausible "random" models involve a variety of biological assumptions about the nature of the species involved, their vagility, colonization processes, population growth, the carrying capacity of the environment for individuals and species, and related phenomena (Grant and Abbott 1980; Colwell and Winkler 1984). As with any other contributing causes, it is infinitely unlikely that these processes explain all of the observed variation in nature, and thus the "null hypotheses" merely represent some of the many feasible determinants of community structure whose relative contributions might be assessed. The value of "null models" is certainly as a construct from which departures may be measured to estimate the impact of processses (e.g., competition) not embodied in the model. The reliability of such estimates, however, depends upon being able to state the model explicitly and estimate its parameters more accurately than those of the process being evaluated. Null hypotheses in ecology are often unsatisfactory because they are virtually impossible to specify completely, or require knowledge unavailable directly and difficult to estimate independently of the pattern being studied.

We will develop these themes further using examples from published studies revolving around a theme of hypothesis testing. Our object is not to criticize the particular findings of the studies chosen, which as a whole represent laudable attempts to systematize and organize an often diffuse and nonrigorous literature, but rather to illustrate that rigid adherence to a logical formalism of testing and rejecting (or failing to reject) supposedly alternative explanations can easily lead to rigorously fallacious conclusions.

NON-ALTERNATIVE HYPOTHESES—MECHANISMS OF PLANT SUCCESSION

Plant succession was originally viewed as a process in which earlier colonists change the physical conditions in a newly available or disturbed habitat, thereby rendering it suitable for later colonists who could not have survived the earlier conditions. In the process these later colonists change conditions so that the early ones cannot persist. In its extreme form, succession was viewed as analogous to ontogeny, with earlier stages (species) being necessary to pave the way for later stages (Clements 1916, 1928, 1936). From the beginning this view was challenged by Gleason (1917, 1926, 1927) and others, who felt that the paucity of late successional species in recently disturbed sites represented slow dispersal and growth, rather than any general need to have the habitat modified by earlier colonists. (See Drury and Nisbet [1973] for a review.)

More recently, Connell and Slatyer (1977) proposed that successional patterns may be characterized as belonging to one of three alternative types, which they term "facilitation," "tolerance," and "inhibition." Facilitation represents the classical view of succession in which early colonists are required for, or increase the rate of establishment of later species. Inhibition occurs when early species retard the arrival or establishment of later species. Tolerance is more akin to a Gleasonian view of succession, in which all species can invade immediately following a disturbance. Late successional and climax species come to predominate because they persist longer than early species, and slowly replace them as they die or are removed by local disturbance.

The three successional models are presented as alternative in the sense that properly designed experiments can in principle distinguish which one actually describes any particular situation (Connell and Slatyer 1977). A sufficient test then is whether late successional species become established more rapidly, less rapidly, or equally rapidly, relative to a control, in a plot in which an early species has been thinned or removed (Sousa 1979). Clearly if this experiment were performed in an undisturbed community consisting of one successional and one climax species, the possible outcomes are that removal of the successional species will slow (facilitation), accelerate (inhibition), or possibly have exactly no effect (tolerance) on the later species. In this case, the three models are mutually exclusive and exhaustive. Tolerance, however, appears to be a null hypothesis, of measure zero, to the other two, unless some arbitrary set of weak interactions is assumed equivalent to no interaction, as in applied inferential statistics.

In almost any more complex case, experimental results that do not allow one to distinguish between the "alternative" processes are possible, or even probable. Performing the test with two species requires knowledge of which species is "climax" and which "successional." If both species persist over long periods of time, one species accelerates establishment of the second, and the second inhibits establishment of the first, facilitation and inhibition could be said to occur simultaneously. Alternatively, at low densities, one species might facilitate the establishment of the second (e.g., by fixing atmospheric nitrogen), but at high densities inhibit it (perhaps by shading). The problems become further magnified when more than two species are considered. In multispecies successions it is generally impossible to rank species exactly in competitive ability or order of appearance (Quinn 1979, 1982, in prep.); thus ambiguities must be present in the model and experimental procedure. Cases of intransitive competitive relationships (A displaces B, B displaces C, but C displaces A; Buss and Jackson 1979) seem particularly difficult to incorporate into Connell and Slatyer's formalism. Problems of density-dependent interactions are magnified by the possibility of interaction (e.g., A and B individually facilitate C, but a mixed stand of A and B inhibits the establishment of C). Finally, it seems likely in any complex succession, that some early colonists (e.g., nitrogen fixers) will facilitate later invaders, whereas other early colonists, if established, will slow the arrival of later invaders. In this case, assignment of the appropriate "alternative" hypothesis will depend upon the successional species chosen for experimentation.

The inability of three simple characterizations to encompass the full diversity of successional processes should be neither surprising nor disturbing, and in some circumstances these simple models may be quite useful (Connell and Slatyer 1977; Usher 1979; Quinn 1979; Sousa 1979). Models of succession, however, illustrate the general problem that nontrivial simple hypotheses about complex systems are never exactly correct. In general, as detailed knowledge of natural history increases, cases that do not fit any of the hypotheses directly, and contain major elements of several, are sure to arise. In such cases, understanding may be better served by direct communication of results of observations and experiments designed to measure underlying processes, rather than by forcing observations into the form of a set of artificially distinct "hypotheses."

INTERACTION BETWEEN HYPOTHESES—DISTRIBUTIONS OF INTERTIDAL ORGANISMS

Ecology textbooks are replete with studies of the causes of distributional limits of species (Krebs 1978; Whittaker 1975). Some of the classical studies come from intertidal communities, in which abrupt limits in vertical distribution of species are often observed (Lewis 1964; Stephenson and Stephenson 1972; Ricketts and

Calvin 1968; Carefoot 1977). Connell (1961a, 1961b, 1970) demonstrated that the distributions of barnacles on rocky shores could be experimentally modified by removing competitors or predators. His suggestion that upper limits to distributions are limited by physical tolerances (e.g., desiccation, freezing, or heat) and lower limits by biological interactions (e.g., predation and competition) is now well established. For any given species, this suggests a series of critical experiments to distinguish between major hypotheses for the causes of distributional limits. Frequently considered hypotheses include: (1) The limits represent physiological limits for survival. (2) The species does not recruit into the area. (3) Limits are set by the action of a competitor. (4) Limits are set by predation.

The hypothesis that a species is not found outside its adult range because of failure to disperse or settle into the area may be rejected by showing that recruits settle into the area (Connell 1961b). This could be done by providing free substrate, e.g., a cleared caged plot or settling plate, and by finding natural recruits. Physical limitation could be eliminated by showing that individuals survive after being transplanted outside their normal range and given appropriate protection from other species (perhaps by caging). Limitation by a "biological enemy," such as a competitor or predator, would be indicated if removal of the second species led to an increase in range, and rejected if it led to a decrease or did not change.

Clearly the hypotheses posed above are not of necessity mutually exclusive. It would be quite possible to have distributional limits affected simultaneously by increasing physical stress, competition, and predation. A more subtle problem emerges when actual experimental results are examined. In most studies, physical stress does not appear to determine the lower limits of intertidal distributions (reviewed by Connell 1972; Carefoot 1977). Yet experiments frequently produce a pattern that would appear to lead to rejection of a hypothesis of limitation by a second species. When a generalized predator is removed, distributions of many prey species typically contract as they are displaced by one or a few dominant competitors. This has been termed a "keystone predator" effect (Paine 1969), and has been repeatedly observed in the field and produced in controlled experiments (Paine 1966, 1971; Harper 1969; Paine and Vadas 1969; Connell 1971, 1978; Dayton 1971; Lubchenco 1978; among others). Even though these observations would appear to reject biological limitation hypotheses, the experimental outcome of the "critical experiment" on predators clearly results from an interaction of predation and competition, and produces an effect opposite to that predicted from the action of either alone (e.g., removal of a "biological enemy" results in contraction, rather than expansion of range).

Analogous results have been found for interactions between physical stress and competition (Dayton 1971; Levin and Paine 1974; Connell 1978; Sousa 1979; Paine and Levin 1981). It seems a common characteristic of marine benthic communities that physical factors causing chronic mortality actually contribute to the persistence of many species by simultaneously removing their principal competitors. Paine (1979) has found that three-way interactions between competition, physical disturbance, and dispersal ability are needed to explain the persistence of a common seaweed. Similarly, physical disturbance in the form of wave shear often appears to decrease predation rates, diminishing the keystone predator effect and increasing the probability of competitive exclusion (Quinn 1979).

In all of these cases, much of the convincing evidence for the roles of, and interactions between, causative factors comes from controlled experiments. Although frequently undertaken with the intent of falsifying the hypothesis that a predator causes prey distributions to be as they are, the principal value of a predator removal experiment seems to be as a relatively direct measure of the effect of the predator on prey distributions. Experiments to clarify ecological causality would seem to be more appropriately directed toward measuring the impact of, and interaction with, potential influences, rather than somehow eliminating all but one by experiment.

"NULL" HYPOTHESES

In order to safely estimate the impact of a particular causative factor, it is necessary to examine two situations in which the strength of the putative cause differs. This is frequently accomplished by comparing situations in which the factor operates with ones in which it is absent. In experiments, controls allow this comparison, but in systems not amenable to experimental manipulation, appropriately constructed null models are used to describe the system in the absence of the action of the postulated causal process.

Perhaps the most troublesome applications of null model techniques in ecology involve attempts to study the importance of interspecific interactions, particularly competition, in determining species' distributions and abundances. Effects of interspecific interaction are notoriously difficult to measure directly without recourse to experimental techniques, which in many cases are infeasible. Natural communities without interspecific interaction do not exist, and thus may not be called upon as reference points for comparison. "Null models" are therefore used to mimic the behavior of hypothetical noninteractive communities.

Null models are in no sense uniquely defined for any natural setting. There seems little possibility species numbers and distributions can generally be deduced from first principles (but see Caswell 1976). More biological reality may be incorporated in the form of species to be considered, physical and spatial limitations on species movement, survival, population growth, and interspecific differences in mortality, resource use, and habitat preference. The choice of such factors considered may drastically alter the predicted nature of the "null" community. The precision and accuracy with which comparisons can be made depends on the reliability of estimates of biological parameters of the "null" models. In many models, estimates cannot be made independently of the actual distributions to which predictions of the model will be compared, and statistical inferences about the predictive power of "null" models may be extremely problematical.

As an example of proposed applications of null models, we will examine studies of the distribution of animals on islands, but many of the difficulties we will discuss apply more generally to the problems of estimating the behavior of any null hypothesis of no interaction in nature.

One approach to studying community dynamics has been to take advantage of "natural experiments," or partially isolated and replicated communities, such as biotas of islands in archipelagoes. Observed patterns are compared with those predicted by particular models or theories. This technique has been used extensively to explore the role of interspecific competition in structuring island vertebrate communities (Diamond 1970, 1978; Grant 1966, 1968, 1969; Schoener 1974, 1975). Agreement between prediction and observation has been taken as support for the applicability of theories of exploitative competition and niche differentiation. This approach has been criticized severely for failure to first test the "null hypothesis" of a noncompetitive community randomly assembled from available colonists to the islands. The critics insist that no interaction may be inferred from distributional data unless the noninteractive case can be rejected (Connor and Simberloff 1979; Strong et al. 1979).

In almost all cases, the hypothesis that island biotas represent random subsamples of the potential colonists that arrived on the islands in the past is posed as an alternative to an important role for competition. (It is rarely specified whether the arrivals of interest are individuals or species.) Unfortunately the actual arrivals are not known in any cases of interest and must be inferred from present distributional data of some kind. In the case of island biotas, reliable estimation of the expected distribution of colonists in the absence of interspecific interaction on the islands requires a number of biological assumptions that seem no more compelling than that of competition.

1. The species pool sampled must be taxonomically appropriate. Grant and Abbott (1980) have observed that the power to detect interspecific interaction declines as more distantly related taxa, presumably less likely to have a major impact on one another, are included in the species pool. Support for this contention comes from Connor and Simberloff's (1979) analysis of the occurrence of pairs and trios of bird or bat species in the West Indies. Exclusive pairs or trios of species within the same family occur 37%–56% more frequently than predicted by the "null hypothesis," whereas exclusive pairs and trios chosen without regard to family never show a deviation of more than 7%. In all of the West Indies comparisons presented, exclusive groupings appear more frequently than predicted by Connor and Simberloff's particular model of "chance." This is consistent with the qualitative predictions of competitive structuring, but the apparent strength of the competitive effect would appear to be stronger within a family. Processes other than competition may, of course, also be consistent with these observations (Simberloff and Connor 1981).

2. The geographic source of potential colonists must be specified to estimate the source pool. How this is done will depend on the particular model. In cases where colonists are assumed to come from a mainland source (MacArthur and Wilson 1967; Simberloff 1974), the number of species, and their relative abundances and probabilities of reaching an island will all depend upon the extent of the presumed source area. Larger areas will place more species in the estimated colonist pool, but more distant individuals or species will have lower expected arrival rates. In actuality, the potential source areas are likely to vary from species to species, not necessarily independently of the competitive processes to be tested against the "non-competitive" colonization model. The source estimation is further confounded if colonization occurs between islands, as the source may not be estimated independently of the distribution of species used in the test.

3. The probability of arrival and establishment must be known for each species.

The implicit assumption in many simple models (e.g., MacArthur and Wilson 1967) that each species in the pool is equally likely to invade is certainly not strictly true, and for many purposes does not provide even a useful approximation. However, the actual probability of invasion and persistence, even in the absence of interspecific interaction, is a function of the probability, or rate, of arrival, the probability of increase in numbers once present, and the probability of extinction following establishment. These probabilities will vary with abundance and distribution within the source area, vagility, distance from the source, birth and death rates under the particular (presumably variable) physical conditions and resource levels on each island, and in the case of many active dispersers, such as birds, individual choice. Under most conditions, few of these parameters are likely to be known much more reliably than the intensity of competition. A methodology which requires their use to establish the reality of interspecific interaction seems destined to failure.

One way of dealing with some of these estimation problems has been to estimate the probabilities of arrival and establishment of species from the proportion of islands in any size class occupied by a particular species (the "incidence function" of Diamond [1975]; Connor and Simberloff 1979; Simberloff and Connor 1981). This technique, however, yields predictions of species distributions derived from the distributions used to test the predictions, so the test is in no way independent. In particular, post-colonization competitive exclusion will be incorporated into the estimate of noninteractive colonization rates, artificially improving the fit to the noninteractive model (Colwell and Winkler 1984; Diamond and Gilpin 1982).

4. Parameters of the noninteractive model must not be estimated from an interactive biota. In some cases, such as the examination of bill-size ratios in insular birds, the species interaction hypothesis is that the patterns observed are those allowed by the predicted competitive interactions, e.g., some limiting similarity and even spacing of sizes (Schoener 1965, 1974; Grant 1968; Abbott et al. 1977). The corresponding null hypothesis is that the distribution of the character is that which would arise in the absence of interaction. There seems no a priori basis for choosing such a distribution. Clearly not all bill sizes are equally probable, and those observed may be the result of a variety of unknown historical events, including competition. One estimation procedure that has been attempted has been to choose species randomly from the species list of a presumed source area and to use the character distributions from those samples as the "null" distribution (e.g., Strong et al. 1979; Schoener 1984). However the degree to which these samples mimic a noninteractive pattern surely depends upon an assumption that the source pool is essentially noninteractive. Subsampling from a fauna highly structured by competition no doubt yields samples showing considerable competitive structure. For example, if there were a "limiting similarity" principle operating in the source pool, there could be no species more similar than the limiting amount in even the most noncompetitive derived fauna (see also Colwell and Winkler 1984). Yet it is clearly unreasonable to require that the role of competition in a complex source fauna be understood in order to permit study of competition in a simple island fauna.

The bottom line is that, no matter how heuristically desirable it may seem, measuring the impact of biological interaction against the reference point of a noninteractive null hypothesis is often not a realistically achievable goal. The characteristics of a noninteractive biota are not known from first principles, and cannot be empirically measured. Estimation of its characteristics depends upon knowledge of other biological parameters, such as distribution and abundance patterns in the source area, dispersal distances, detailed colonization processes, etc., that are no better established than the consequences of interspecific interaction. In practice, the purported null hypotheses are better viewed as dispersal models, alternative causes in the non-mutually-exclusive sense discussed above. Viewed this way, "random" or "null" models have no "logical primacy" over other possible causative factors in the sense claimed by Strong (1980). On the other hand, we agree with Simberloff and Connor (1981) that the ease of constructing noncompetitive models which predict patterns similar to those ostensibly resulting from competition makes the measurement of competitive effects from distributional data unreliable, and a conclusion of strong competitive effects uncompelling in most cases. When feasible, direct experimentation is desirable (see also Connell 1975; Dunham 1980; Grant 1972; Hairston 1981; Paine 1966, 1971). We find Simberloff and Connor's claim that a random colonization model is more "parsimonious" than a competition model to be distinctly a matter of taste.

DISCUSSION

Throughout this paper, we have argued that a strict application of a formal "strong inference" methodology to elucidating potential causes of patterns in nature is frequently infeasible. Putative causes generally cannot be stated in a way that they are either mutually exclusive or potentially global in their application. Critical experiments to distinguish between the truth of "alternative" causes cannot be performed in principle, and the criteria for, and logical import of, falsification of a potential cause is unclear. We believe that rigid insistence on the "hypothesis testing" formalism has the potential to distract from understanding in several ways. Treating useful generalizations, such as models of succession, as well-defined alternatives denies the possible richness of a continuous range of possible outcomes and suggests inappropriate experiments to distinguish between single points in that range, each infinitely improbable as a description of the actual truth. Non-alternative causes may interact and influence patterns observed in a way that will not be detected by the kind of univariate critical experiment that would be used to attempt to reject proper Popperian mutually exclusive hypotheses. As illustrated by the discussion of a simple interaction in the intertidal zone, such experiments can even lead to erroneous rejection of an important causal process.

Processes contributing to pattern in natural communities do not often lend themselves to easy statement as hypotheses of hypothetico-deductive formalism, since relative contribution and possible interaction are the objects of investigation, not truth or falsity of the process. Ostensible "critical tests" often have value in measuring these contributions. "Rejection" sets the statistical limits of detection as probable upper bounds to the estimate of the process's relative impact. Thus the role of the null hypothesis is as a reference point for measurement of unknown departures, in the spirit of statistical hypothesis testing, rather than as an alternative hypothesis with some probability of being strictly true, in the sense of "strong inference." "Null" models are not the only possible reference points. For example, optimality models may serve a similar role, also with no probability of representing strict truth.

In other cases, logical reference points may not exist. For example, in a study of temperature effects, a null temperature analogous to a null model of no competition is difficult to define. Most investigators would proceed, if possible, to manipulate temperature experimentally, and measure its effect by regression. By the same logic, there seems no logical imperative that a null model be considered (or as we have argued, even be usefully definable) in an ecological or evolutionary investigation.

We do not intend, in any of our discussion, to downplay either the value of considering multiple approaches to biological problems, or the dangers of attempting to demonstrate preconceived causality without adequate consideration of other potentially contributing processes (Chamberlin 1897; but see Campbell 1979). We believe, as do most of the outspoken advocates of formal hypothesis testing, that controlled experimentation, when feasible, is a powerful technique for removing uncertainty from our understanding of the natural world. We suggest, however, that ecology and evolution are not blessed with clearcut criteria for acceptance of theories, much less methodological prescriptions or requirements for successful science. We view the changes of understanding in these fields as perhaps more akin to Kuhn's model of the establishment of paradigms in science than to Platt's model of strong inference, although we see no reason to distinguish between large scale changes in viewpoint (if there have been any since Darwin) and more modest theoretical advances. Theories are embraced when, in part, a relatively simple explanation seems to account satisfactorily for much of a complex set of observations, and are abandoned or modified as the weight of post hoc additions becomes a burden, and other, comparably simple and appealing viewpoints are suggested. Consideration of alternatives and careful experimentation obviously contribute to this process. Formal method is a guide to innovation, however, not a requirement, and healthy skepticism toward a single methodological model seems thoroughly as appropriate as toward any other claim of scientific truth.

SUMMARY

Theories of causality in ecology and evolution rarely lend themselves to analysis by the formal method of "hypothesis testing" envisioned by champions of a "strong inference" model of scientific method. The objective of biological research typically is to assess the relative contributions of a number of potential causal agents operating simultaneously. Sensibly stated hypotheses in the methodology of most field investigations are similar to hypotheses of applied statistics. They are not intended to be mutually exclusive, in any sense exhaustive, or global in their application. It is not possible in principle to perform a "critical test" or experiment to distinguish between the truth of "alternative hypotheses" if the proposed causal processes they caricature occur simultaneously.

We consider several examples in which a rigid hypothetico-deductive methodology applied to nonalternative ecological "hypotheses" could lead to fallacious conclusions. It has been proposed that processes of ecological succession may be separated into alternative modes of "facilitation," "inhibition," and "tolerance." Yet attempts to experimentally reject one or more of the supposedly distinct hypotheses cannot, in principle, distinguish between them in a variety of biologically interesting cases. In studies of the limits of distributions of intertidal organisms, reasonable univariate experimental tests of possible causes would lead to rejection of "biological enemy" hypotheses when a "keystone predator effect" occurs because the interaction between competition and predation reverses the direction of the effect on some prey populations expected from either process in isolation.

Particular problems arise when "null models" in ecology are treated as hypotheses of "strong inference." Models of ecological or evolutionary causality rarely have single or easily stated "null" converses. Tractable null models have no probability of being strictly true, and thus may be rejected a priori as hypothetico-deductive constructs. In practice, their role is as a reference point for measurement of departures. Their usefulness in this regard depends upon the reliability with which the characteristics of biology without interaction can be estimated. Applied to studies of interspecific competition through the use of species distributions, purported null hypotheses make different biological assumptions than those of the interactive models. They seem neither especially more reliable nor in any way more fundamental. We see no reason to accept the recent claims that "null hypotheses," as applied in ecology and evolution, have any logical primacy or greater parsimony than other approaches to partitioning the variation observed in natural communities among the contributions of many observable causes.

Careful consideration of possible explanations and controlled experimentation contribute a great deal to ecological and evolutionary knowledge. However, we believe that the hypothetico-deductive model of scientific method can provide misleading prescriptions for efficient investigation and acceptance of evidence in phenomena with multiple causes, and should be applied with appropriate skepticism.

ACKNOWLEDGMENTS

We thank J. Connell, R. Davis, P. Grant, K. Hopper, R. Karban, B. Milligan, P. Richerson, D. Schluter, T. Schoener, and C. Toft for valuable comments on earlier drafts.

LITERATURE CITED

Abbott, I., L. K. Abbott, and P. R. Grant. 1977. Comparative ecology of Galapagos ground finches: evaluation of the importance of floristic diversity and interspecific competition. Ecol. Monogr. 47:151–184.

Bacon, F. 1620. Novum organum. Part 2. Instauratio magna. J. Billium, London.
Brush, S. 1974. Should the history of science be X-rated? Science 183:1164–1172.
Buss, L. W., and J. B. C. Jackson. 1979. Competitive networks: nontransitive competitive relationships in cryptic coral reef environments. Am. Nat. 113:223–234.
Campbell, D. T. 1979. A tribal model of the social system vehicle carrying scientific knowledge. Knowledge: Creation, Diffusion, Utilization 1:181–201.
Carefoot, T. 1977. Pacific seashores. University of Washington Press, Seattle.
Caswell, H. 1976. Community structure: a neutral model analysis. Ecol. Monogr. 46:327–354.
Chamberlin, T. C. 1897. The method of multiple working hypotheses. J. Geol. 5:837–848.
Clements, F. E. 1916. Plant succession. Carnegie Inst. Wash. Publ. 242.
———. 1928. Plant succession and indicators. H. W. Wilson, New York.
———. 1936. Nature and structure of the climax. J. Ecol. 24:252–284.
Colwell, R. K., and D. W. Winkler. 1984. A null model for null models in biogeography. *In* D. S. Strong, Jr., D. S. Simberloff, L. G. Abele, and A. B. Thistle, eds. Ecological communities: conceptual issues and the evidence. Princeton University Press, Princeton, N.J. (in press).
Connell, J. H. 1961a. Effects of competition, predation by *Thais lapillus* and other factors on populations of the barnacle *Balanus balanoides*. Ecol. Monogr. 31:61–104.
———. 1961b. The influence of interspecific competition and other factors on the distribution of the barnacle *Chthamalus stellatus*. Ecology 42:710–723.
———. 1970. A predator-prey system in the marine intertidal region. I. *Balanus glandula* and several predatory species of *Thais*. Ecol. Monogr. 40:49–78.
———. 1971. On the role of natural enemies in preventing competitive exclusion in some marine animals and in rain forest trees. Pages 298–312 *in* P. J. den Boer and G. Gradwell, eds. Dynamics of numbers in populations. Proceedings of the Advanced Study Institute on dynamics of numbers in populations, Osterbeek, 1970. Centre for Agricultural Publishing and Documentation, Wageningen.
———. 1972. Community interactions on marine rocky intertidal shores. Annu. Rev. Ecol. Syst. 3:169–192.
———. 1975. Some mechanisms producing structure in natural communities: a model and evidence from field experiments. Pages 460–490 *in* M. Cody and J. Diamond, eds. Ecology and evolution of communities. Harvard University Press, Cambridge, Mass.
———. 1978. Diversity in tropical rain forests and coral reefs. Science 199:1302–1310.
Connell, J. H., and R. O. Slatyer. 1977. Mechanisms of succession in natural communities and their role in community stability and organization. Am. Nat. 111:1119–1144.
Connor, E. F., and D. Simberloff. 1978. Species number and compositional similarity of the Galapagos flora and avifauna. Ecol. Monogr. 48:219–248.
———. 1979. Assembly of species communities: chance or competition? Ecology 60:1132–1140.
Dayton, P. K. 1971. Competition, disturbance, and community organization: the provision and subsequent utilization of space in a rocky intertidal community. Ecol. Monogr. 41:351–389.
Diamond, J. M. 1970. Ecological consequences of island colonization by southwest Pacific birds 1: types of niche shifts. Proc. Natl. Acad. Sci. USA 67:529–536.
———. 1975. Assembly of species communities. Pages 342–444 *in* M. L. Cody and J. M. Diamond, eds. Ecology and evolution of communities. Harvard University Press, Cambridge, Mass.
———. 1978. Niche shifts and the rediscovery of interspecific competition. Am. Sci. 66:322–331.
Diamond, J. M., and M. Gilpin. 1982. Examination of the "null" model of Connor and Simberloff for species co-occurrences on islands. Oecologia 52:64–74.
Drury, W. H., and I. C. T. Nisbet. 1973. Succession. J. Arnold Arbor. 54:331–368.
Dunham, A. E. 1980. An experimental study of interspecific competition between the iguanid lizards *Sceloporus merriami* and *Urosaurus ornatus*. Ecol. Monogr. 50:309–330.
Eiseley, L. 1973. The man who saw through time. Scribner's, New York.
Feyerabend, P. K. 1975. Against method. Humanities Press, London.
Gleason, H. A. 1917. The structure and development of the plant association. Bull. Torrey Bot. Club 44:463–481.
———. 1926. The individualistic concept of the plant association. Bull. Torrey Bot. Club 53:7–26.
———. 1927. Further views on the succession concept. Ecology 8:299–326.

Grant, P. R. 1966. Ecological incompatibility of bird species on islands. Am. Nat. 100:451–462.
———. 1968. Bill size, body size, and the ecological adaptations of bird species to competitive situations on islands. Syst. Zool. 17:319–333.
———. 1969. Colonization of islands by ecologically dissimilar species of birds. Can. J. Zool. 47:41–43.
———. 1972. Convergent and divergent character displacement. J. Linn. Soc. 4:39–68.
Grant, P. R., and I. Abbott. 1980. Interspecific competition, island biogeography, and null hypotheses. Evolution 34:332–343.
Hacking, I. 1965. The logic of statistical inference. Cambridge University Press, Cambridge.
Hairston, N. J. 1981. An experimental test of a guild: salamander competition. Ecology 62:65–72.
Harper, J. L. 1969. The role of predation in vegetational diversity. Brookhaven Symp. Biol. 22:48–61.
Krebs, C. J. 1978. Ecology: the experimental analysis of distribution and abundance. 2d ed. Harper & Row, New York.
Kuhn, T. S. 1970. The structure of scientific revolutions. University of Chicago Press, Chicago.
Lakatos, I. 1970. Falsification and the methodology of research programmes. Pages 91–195 *in* I. Lakatos and A. Musgrave, eds. Criticism and the growth of knowledge. Cambridge University Press, Cambridge.
———. 1974. Popper on demarkation and induction. Pages 241–273 *in* P. Schilpp, ed. The philosophy of Karl Popper. Open Court, LaSalle, Ill.
Lawlor, L. R. 1980. Structure and stability in natural and randomly constructed competitive communities. Am. Nat. 116:394–408.
Levin, S. A., and R. T. Paine. 1974. Disturbance, patch formation and community structure. Proc. Natl. Acad. Sci. USA 71:2744–2747.
Levins, R., and R. Lewontin. 1980. Dialectics and reductionism in ecology. Synthèse 43:47–78.
Lewis, J. R. 1964. The ecology of rocky shores. English University Press, London.
Lubchenco, J. 1978. Plant species diversity in a marine intertidal community: importance of herbivore food preference and algal competitive abilities. Am. Nat. 112:23–39.
MacArthur R. H., and E. O. Wilson. 1967. The theory of island biogeography. Princeton University Press, Princeton, N.J.
Paine, R. T. 1966. Food web complexity and species diversity. Am. Nat. 100:65–75.
———. 1969. A note on trophic complexity and community stability. Am. Nat. 103:91–93.
———. 1971. A short-term experimental investigation of resource partitioning in a New Zealand rocky intertidal habitat. Ecology 52:1096–1106.
———. 1979. Disaster, catastrophe, and local persistence of the sea palm *Postelsia palmaeformis*. Science 205:685–687.
Paine, R. T., and S. A. Levin. 1981. Intertidal landscapes: disturbance and the dynamics of pattern. Ecol. Monogr. 51:145–178.
Paine, R. T., and R. L. Vadas. 1969. The effect of grazing by sea urchins *Strongylocentrotus* spp. on benthic algal populations. Limnol. Oceanogr. 14:710–719.
Peters, R. H. 1976. Tautology in evolution and ecology. Am. Nat. 110:1–12.
Platt, J. R. 1964. Strong inference. Science 146:347–353.
Popper, K. R. 1959. The logic of scientific discovery. Basic Books, New York.
———. 1972. Objective knowledge: an evolutionary approach. Clarendon, Oxford.
———. 1976. Unended quest: an intellectual autobiography. Open Court, La Salle, Ill.
———. 1983*a*. A proof of the impossibility of inductive probability. Nature 302:687–688.
———. 1983*b*. Realism and the aim of science. Vol. I of the postscript to the logic of scientific discovery. Hutchinson/Rowan & Littlefield, London.
Quinn, J. F. 1979. Disturbance, predation and diversity in the rocky intertidal zone. Ph. D. diss. University of Washington, Seattle.
———. 1982. Competitive hierarchies in marine benthic communities. Oecologia 54:129–135.
Ricketts, E. F., and J. Calvin. 1968. Between Pacific tides. 4th ed. Stanford University Press, Stanford, Calif.
Schoener, T. W. 1965. The evolution of bill size differences among congeneric species of birds. Evolution 19:189–213.
———. 1974. Resource partitioning in ecological communities. Science 185:27–39.

———. 1975. Presence and absence of habitat shift in some widespread lizard species. Ecol. Monogr. 233–258.
———. 1984. Size differences among sympatric, bird-eating hawks: a worldwide survey. *In* D. S. Strong, Jr., D. S. Simberloff, L. G. Abele, and A. B. Thistle, eds. Ecological communities; conceptual issues and the evidence. Princeton University Press, Princeton, N.J. (in press).
Simberloff, D. S. 1974. Equilibrium theory of island biogeography and ecology. Annu. Rev. Ecol. Syst. 5:161–182.
———. 1978. Using biogeographic distributions to determine if colonization is stochastic. Am. Nat. 112:723–726.
———. 1980. A succession of paradigms in ecology: essentialism to materialism and probabilism. Synthèse 43:3–39.
Simberloff, D. S., and E. F. Connor. 1981. Missing species combinations. Am. Nat. 118:215–239.
Stephenson, T. A., and A. Stephenson. 1972. Life between tidemarks on rocky shores. Freeman, San Francisco.
Strong, D. R., Jr. 1980. Null hypotheses in ecology. Synthèse 43:271–285.
Strong, D. R., Jr., L. A. Szyska, and D. S. Simberloff. 1979. Tests of community-wide character displacement against null hypotheses. Evolution 33:897–913.
Sousa, W. P. 1979. Experimental investigations of disturbance and ecological succession in a rocky intertidal community. Ecol. Monogr. 49:227–254.
Usher, M. B. 1979. Markovian approaches to ecological succession. J. Anim. Ecol. 48:413–426.
Whittaker, R. H. 1975. Communities and ecosystems. Macmillan, New York.

DETECTING COMMUNITY-WIDE PATTERNS: ESTIMATING POWER STRENGTHENS STATISTICAL INFERENCE

CATHERINE A. TOFT AND PATRICK J. SHEA

Department of Zoology, University of California, Davis, California 95616; U.S.D.A. Forest Service, Pacific Southwest Forest and Range Experiment Station, P.O. Box 245, Berkeley, California 94701

Submitted June 29, 1982; Accepted October 21, 1982

Ecologists are now regularly evaluating null models to look for mechanisms (usually competition) that produce community-wide patterns (see Strong et al. 1984 for the current state of the art). A number of investigators have failed to reject such null hypotheses, concluding either that the mechanism producing a particular pattern is absent or, more conservatively, that a certain pattern has not yet been statistically demonstrated to exist (e.g., Connor and Simberloff 1979; Strong et al. 1979; Simberloff and Boecklen 1981; Simberloff and Connor 1981; Roth 1981; Wiens and Rotenberry 1980, 1981). None of these papers has mentioned the statistical power of tests that fail to reject null hypotheses (but see Colwell and Winkler 1984); power may be thought of as the ability of a test to determine whether the null hypothesis is false. Here we make some simple yet important points about the use of power analysis in statistical inference as it affects ecological hypotheses. We conclude that estimation of power constitutes a major statistical tool for use in discerning ecological patterns—one that is vastly underexploited in ecological research.

TYPE I AND TYPE II ERRORS

Null hypotheses in ecology have been the subject of much recent discussion (papers cited above; also Grant and Abbott 1980; Case and Sidell 1983; Diamond and Gilpin 1982; Quinn and Dunham 1983; Roughgarden 1983). In ecology, perhaps appropriately or even inevitably (Quinn and Dunham 1983), hypotheses are typically statistical hypotheses and not true Popperian ones. Unlike the latter which are supposed to be mutually exclusive and either true or false, tests of statistical hypotheses are probabilistic, and statistical tests can just as easily estimate the degree to which an effect is felt as judge whether it exists or not (Cohen 1977). As a result, we may compute among other things the probability of two types of error: type I error (α) or the probability that we have mistakenly rejected a true null hypothesis; and type II error (β) or the probability that we have mistakenly failed to reject a false null hypothesis (table 1). The power of a

TABLE 1

Decision Table for Hypothesis Testing and Associated Probabilities

"Truth"	Decision	
	Accept H_0	Reject H_0
H_0 True	No Error $1 - \alpha$	Type I Error α
H_0 False	Type II Error β	No Error $1 - \beta$

test is simply the probability of not committing type II error, that is, $1 - \beta$. The more powerful a test, the more likely it is to show, statistically, an effect that exists.

We are most familiar with the pitfalls of type I error and are, and have been for some time, obligated to publish α's for all statistical tests that we do. Accordingly, there is a critical level of α of .05 which is a widespread, almost inviolate, convention. In contrast, type II error is rarely estimated in basic research, and there is no single critical level of power in widespread use. The reason is quite understandable and is not entirely that power tables for most statistical analyses have been unavailable until rather recently. Scientists doing basic research are by nature cautious and rightly so. An effect that is real will eventually, after sufficient inquiry, manifest itself; to jump prematurely to a conclusion that turns out to be false is by far a worse error than failing to detect something. In other words, the cost of type I error most often greatly exceeds the cost of type II error in basic research.

In ecological research, whether basic or applied, there may be situations in which knowing the power of a test, i.e., the probability of type II error, is useful, perhaps even vital. One situation is when type II error has a substantial cost. Another is when statistical inference is used to measure the degree of an effect. That is, if we know that the probability of committing type II error is low, the degree of departure from the null hypothesis measures the degree to which an effect occurs for a given amount of confidence. After we present some necessary basics about power analysis, we give examples of both situations and then present some general recommendations about the use of power analysis in ecological research.

SOME DETAILS ABOUT POWER ANALYSIS

Power depends on three factors, mostly determined a priori by the investigator (Cohen 1977). The first is the critical level of α used. Quite obviously, the more stringent one's standards for avoiding type I error, the more likely it is that one will commit type II error. For example (table 2), suppose an investigator does a chi-square test, as is common in the ecological literature. If he decides to set an

TABLE 2

RELATIVE COSTS OF TYPE I AND TYPE II ERRORS FOR DIFFERENT α†

Critical α	.05	.01	.001
Power‡ $(1 - \beta)$	.85	.66	.37
Relative cost (ratio of type II to type I error)	3	34	630

† For contingency tables where df = 1, N = 100, and w(effect size) = .30 (medium, Cohen 1977).
‡ Determined from tables in Cohen (1977) for α = .05, α = .01, and Haynam et al. (1970) for α = .001.

extremely stringent critical value of α = .001, the power of that test will be relatively low. Under the fairly representative conditions in table 2, the ratio of type II to type I error (.63/.001) tells us that this investigator considers type I error to be over 600 times more costly than type II error! As he "lowers" his standards and sets more liberal α's, the power of his test will go up accordingly. With an α of .05, the investigator accepts a 3 to 1 ratio of type II to type I error, probably a more reasonable representation of the relative costs of type I and type II errors. Perhaps an extreme ratio of type II to type I errors, such as in the first case, might be indeed satisfactory to an investigator, but few investigators seem aware of the decision they make about the relative costs of the two types of errors.

Second, the larger the sample size, the more reliable the sample estimate and therefore the smaller the probability of both types of errors (Cohen 1977). We are all familiar with standard test statistics, and they are always computed with sample sizes. The investigator may or may not have control over sample sizes in a particular situation; sample size may or may not have been considered a priori. One, perhaps the most familiar, use of power analysis is to determine the sample size necessary to give a certain statistical reliability.

The third consideration is the magnitude of the effect that the investigator wants to be able to detect. This is the so-called "effect size" (Cohen 1977). Intuitively, one can see that the larger the critical effect size, i.e., the larger the difference one is trying to detect or the stronger the effect, the more easily it can be detected and thereby the greater the power in a given test. There is no conventional standard for setting effect size. Rather, this decision must be uniquely tailored to each test and each situation; it simply depends on what the investigator is trying to show. Effect size may be set by theory. A good example in ecology would be the detection of certain d/w values for resource utilization functions (MacArthur and Levins 1967; Roughgarden 1974; May 1975). Where theory does not provide formal guidelines, effect size may alternatively be set by common sense. Cohen (1977) presents guidelines for standardizing effect sizes, thereby making power tests more comparable among studies.

These three pieces of information are all that is required to compute power. Moreover, power tests now can be done easily for many standard statistical analyses. Why then are estimates of power rarely, if ever, provided in ecological papers using such statistical tests? We believe that they are not for some combination of the following reasons: (1) People may be unaware of type II error and its costs. (2) They may be unaware that power tables exist for most common statistical analyses. (3) They may be unaware that power estimates can greatly strengthen some sorts of statistical inference. Below we give examples to address these points.

EXAMPLES

Asserting no difference.—When type II error is conservative, it is rarely costly. In this situation, an investigator states that he has failed to demonstrate an effect. However, some investigators imply or explicitly conclude that the effect does not exist. In the ecological literature, failing to falsify the null hypothesis in some community analyses has led to the conviction that competition does not occur. Perhaps, in the latter case, a positive conclusion from a negative statistical result should be subjected to the same stringent standards as a positive conclusion from a positive statistical result. After all, it is a statement of fact, of proof that a claim has been demonstrated. Here the cost of making a type II error is exactly the same as the cost of making a type I error; this is the cost of making a false claim. It would therefore be reasonable to require, in this situation, that α and β be set at the same critical level. For example, if the critical level of α is .05, we should require a critical power of .95. Of course, we can compute power only if a reasonable, specific alternative, i.e., effect size, can be set.

In applied situations related to management of ecological systems, making a type II error can have more tangible costs, often much greater than costs of type I error. For example, in testing pesticides, the goal is to choose the lowest dosage that will reduce a pest population by a certain amount. If there is no significant difference between the effects of two dosage rates, we choose the lower rate, both to minimize actual costs and to minimize harmful side effects to species that are not pests. If the lack of statistical significance is really caused by unreliability of the test estimates, then when the lower dosage is applied to several thousand or hundred thousand hectares, this error (a type II error) becomes one of great magnitude. It probably can include: the money and time wasted on an ineffective application; the money and time necessary to repeat the application (if it is possible, and it may not be); the economic hardship due to a failed crop; an unnecessary increase in the resistance of the pest to the pesticide; and the loss of credibility by the public toward applied research.

In management of forest pests, the typical experimental design has been to use three replications (plots per treatment) to test insecticides (table 3). In another case similar to the design used in table 3, two experimental dosages achieved about 95% control of western spruce budworm, *Choristoneura occidentalis* Freeman, in a test run; when applied to 8,000 ha, however, the lower dosage achieved only about 50% control. For testing an insecticide aimed at controlling larch casebearer (table 3), three replications provide sufficient power (.80 ± .10) to detect only a 50% or greater difference (the effect size) between two dosage rates, more than encompassing the observed difference between test and actual applications in the case of western spruce budworm. In contrast, 14 replications were necessary to detect reliably a smaller difference of about 10%, one that will

TABLE 3

Sampling Regimes and Estimated Costs for Randomized Block Field Tests of Insecticide Treatments to Larch Casebearer [*Coleophora laricella* (Hübner)] Needlemining Populations (from Hard et al. 1979)

H_A (% difference in mortality between treatments [= effect size])	Treatments (No. of dosage rates of insecticide)	Replications		Experimental Power $1 - \beta$	Minimum Cost per Experiment
		Plots per Treatment	Trees per Plot		
50%	2	3	17	.90	$ 3,590
	2	3	9	.80	2,974
	2	3	6	.71	2,743
10%	2	3	...	>.20	...
	2	14	19	.80	17,474
	2	11	23	.70	14,859
	2	10	18	.60	12,224

provide much greater predictability between the test treatment and the actual application on many thousand hectares. Though the cost of achieving the greater power is much higher (table 3), it is still considerably less than that of committing type II error on a very large scale.

Similarly, in testing drugs, for example, comparing name brands versus generics or detecting absence of harmful side effects (when the goal is to detect no effect), the costs of type II error can be still greater and include human disability and loss of life. While the costs of type II error in basic ecological research are not so dramatic, type II error is perhaps more insidious for that reason: Because our incentive for detecting type II error is lower, we are more likely to be enticed into believing we have discovered some truth when in fact we have committed an error. Though the specific example we present above is from applied research, basic competition experiments have in principle the same design, when adding species and not insecticide constitutes the treatments. What should we conclude when adding (or deleting) a species has no effect in a competition experiment? Here, as in testing insecticides, knowledge of the experimental power is of vital interest and is necessary before definite conclusions can be drawn from the experiment.

Measuring degree.—On more than one occasion, we are sure, audiences have snickered when an investigator has pleaded that $P = .06$ is "almost significant." Such a plea in fact acknowledges a very real and useful aspect of statistical inference: measuring the degree to which an effect occurs (Pielou 1969; Cohen 1977).

The null hypothesis in statistics traditionally states that a particular effect size is zero, or at least so small as to be negligible (Cohen 1977). From zero, or no effect, an effect may be felt in continuously varying degrees until the effect becomes very, very large. In our enthusiasm for avoiding type I error we ignore information provided by test statistics when $P > .05$. When we know the probability of type II error to be small, the degree of departure from the null hypothesis provides information about the strength of an effect both above and below the .05 level of p_α.

An application of current ecological interest can be found in certain tests for competition. Once we have reached a consensus that competition does occur in some systems, the next question concerns the degree to which competition occurs, How does competition vary in nature and what factors cause this variation? Such information will ultimately tell us more about the nature and mechanisms of competition in real communities. When habitats are divided into discrete units, the intensity of competition among species can be statistically estimated using the occurrences of species in those units.

For example, a 2 × 2 contingency test can detect interactions between two species occurring in such units (Pielou 1969; Cohen 1970). The table requires that species' occurrences in each habitat unit be scored in one of four categories: both species present; only species 1 present; only species 2 present; and neither species present. With competition, fewer habitat units than expected contain both species. The null hypothesis, all other things being equal, says that there should be no detectable interaction, i.e., $\chi^2 \simeq 0$. The stronger the degree of competition between two species, the fewer habitat units will contain both species, and χ^2 gets larger. A convenient measure of association (there are others [Pielou 1969; Cohen 1977]) is the contingency coefficient, which is normalized by sample size, N:

$$C = \sqrt{\chi^2/(N + \chi^2)}. \tag{1}$$

As the intensity of competition increases, C approaches (but never reaches) 1, and as the intensity of competition decreases, $C \to 0$. The contingency coefficient C therefore provides a continuously varying measure of the degree to which two species interact. Importantly it is independent of sample size and thus meets Cohen's (1977) requirement for an effect size index: a pure (dimensionless) number that increases with the degree of difference between alternative and null hypotheses, or in this case with the intensity of competition. Only if the power of a test is high enough, however, can C be trusted to reflect the intensity of competition when p_α exceeds .05, rather than to reflect unreliability of the sample data. The power of any given table will depend on the total number of observations, N; N, for a given amount of power, will therefore determine how small an amount of competition (i.e., the effect size) we can reliably detect.

For example, Toft et al. (1982) used this method to measure the relative intensity of competition among five species of subarctic ducks occurring on isolated ponds. There we correctly predicted the relative degree of interaction between two species from the degree of overlap between those species in time and space; we also showed that the intensity of competition varied among years. Thus we were able to learn more about how these species interact than simply finding out whether or not some species competed. With a sample size, N, of 244 habitat units we had sufficient power ($\simeq .9$) in the 2 × 2 contingency tests to detect reliably a moderate amount of interaction (by Cohen's [1977] criteria, i.e., an effect size of $w = .20$). This power was sufficient to rank species-pairs in rough order of competition intensity which suited our purposes in that paper. A sample size of 1,000 or more habitat units in a 2 × 2 table would be necessary to detect small differences in the degree of competition; however, we would recommend more direct methods, such as manipulation, to achieve this degree of resolution.

Our point in this example is simply that knowing the power of a test can greatly strengthen some sorts of statistical inference. If we determine the power in any given test, we can use statistical inference much more effectively to measure the degree to which an effect such as competition occurs and be confident of how accurately we can detect an effect of a certain size. To be able to do so provides more information from a set of data than merely stating whether or not the effect has been shown to occur.

RECOMMENDATION

Where we fail to reject the null hypothesis, it would be valuable to know the reason: Are sample estimates unreliable? Is there no effect of a postulated factor? Or does that factor exert a small or intermediate effect? This information is contained, in part, in the test's power level. This information, moreover, is essential for detecting community-wide patterns, or lack thereof, with statistical models and for inferring mechanisms where such patterns exist. We have given two examples of the use of—in fact, the necessity of—power analysis in ecological inference: asserting no effect and measuring degree. We call on our colleagues to publish power levels of their tests, when possible (but see Colwell and Winkler 1983), or at least provide sufficient information for others to do so. In particular, when strong assertions are based on lack of statistical significance, we should meet the same stringent standards for avoiding type II error as we now do for avoiding type I error.

SUMMARY

Power statistics are drastically underutilized in basic and applied ecological research where they could provide objective measures of the sensitivity of null hypotheses tests and thereby strengthen some statistical inferences. Null models are being used increasingly to investigate the causes of community-wide patterns, yet researchers tend to ignore the risks involved in committing the type II error associated with these models. The three factors that determine power, e.g., critical level of α, sample size, and "effect size" are explained and their effects on power are discussed.

In our examples we have attempted to illustrate how power analysis can assist investigators in interpreting their research results. In ecological studies' failing to demonstrate an effect is quite different than implicitly or explicitly concluding that no difference exists. In this situation, assuming low power, the cost of committing a type II error is that of making false claim. Pesticide and drug examples demonstrate that often there can be serious economic, health, and social costs associated with the commission of type II errors. Lastly we describe a situation whereby power analysis can be used to measure the degree to which an effect occurs and thereby expand our conclusions.

ACKNOWLEDGMENTS

We are grateful to A. P. Fenech, D. H. Johnson, J. F. Quinn, G. W. Salt, T. W. Schoener, and D. Sharpnack for comments on earlier drafts.

LITERATURE CITED

Case, T. J., and R. Sidell. 1983. Pattern and chance in the structure of model and natural communities. Evolution (in press).
Cohen, J. 1977. Statistical power analysis for the behavioral sciences. Academic Press, New York.
Cohen, J. E. 1970. A Markov contingency-table model for replicated Lotka-Volterra systems near equilibrium. Am. Nat. 104:547–560.
Colwell, R. K., and D. W. Winkler. 1984. A null model for null models in biogeography. In D. S. Strong, Jr., D. S. Simberloff, L. G. Abele, and A. B. Thistle, eds. Ecological communities: conceptual issues and the evidence. Princeton University Press, Princeton, N.J.
Connor, E. F., and D. S. Simberloff. 1979. The assembly of species communities: chance or competition? Ecology 60:1132–1140.
Diamond, J. M., and M. E. Gilpin. 1982. Examination of the "null" model of Connor and Simberloff for species co-occurrences on islands. Oecologia 52:64–74.
Grant, P. R., and I. Abbott. 1980. Interspecific competition, island biogeography and null hypotheses. Evolution 34:332–341.
Hard, J. S., S. Meso, and M. Haskett. 1979. Testing aerially applied Orthene for control of larch casebearer. U.S. Dep. Agric. For. Serv. Pac. Southwest For. Range Exp. Stn. Berkeley, Calif. Res. Pap. PSW-138.
Haynam, G. E., Z. Govindarajulu, and F. C. Leone. 1970. Tables of the cumulative non-central chi-square distribution. Pages 1–78 in H. L. Harter and D. B. Owen, eds. Selected tables in mathematical statistics. Vol. 1. Markham, Chicago.
MacArthur, R. H., and R. Levins. 1967. The limiting similarity, convergence, and divergence of coexisting species. Am. Nat. 101:377–385.
May, R. M. 1975. Stability and complexity in model ecosystems. 2d ed. Princeton University Press, Princeton, N.J.
Pielou, E. C. 1969. An introduction to mathematical ecology. Wiley, New York.
Quinn, J. F., and A. E. Dunham. 1983. On hypothesis testing and evolution. Am. Nat. 122:602–617.
Roth, L. V. 1981. Constancy in the size ratios of sympatric species. Am. Nat. 118:394–404.
Roughgarden, J. 1974. Species packing and the competition function with illustrations from coral reef fishes. Theor. Popul. Biol. 58:163–186.
———. 1983. Competition and theory in community ecology. Am. Nat. 122:583–601.
Simberloff, D., and W. Boecklen. 1981. Santa Rosalia reconsidered: size ratios and competition. Evolution 35:1206–1228.
Simberloff, D., and E. F. Connor. 1981. Missing species combinations. Am. Nat. 118:215–234.
Strong, D. R., Jr., D. S. Simberloff, L. G. Abele, and A. B. Thistle, eds. 1984. Ecological communities: conceptual issues and the evidence. Princeton University Press, Princeton, N.J.
Strong, D. R., Jr., L. A. Syzska, and D. S. Simberloff. 1979. Tests of community-wide character displacement against null hypotheses. Evolution 33:897–913.
Toft, C. A., D. L. Trauger, and H. W. Murdy. 1982. Tests for species interactions: breeding phenology and habitat use in subarctic ducks. Am. Nat. 120:586–613.
Wiens, J. A., and J. T. Rotenberry. 1980. Patterns of morphology and ecology in grassland and shrubsteppe bird populations. Ecol. Monogr. 50:287–308.
———. 1981. Habitat associations and community structure of birds in shrubsteppe environments. Ecol. Monogr. 51:21–41.

COMPETITION THEORY, HYPOTHESIS-TESTING, AND OTHER COMMUNITY ECOLOGICAL BUZZWORDS

Daniel Simberloff

Department of Biological Science, Florida State University, Tallahassee, Florida 32306

Submitted October 12, 1982; Accepted April 19, 1983

Roughgarden's (1983) and Quinn and Dunham's (1983) thoughtful essays on how to do community ecological research both fall into two parts: (1) general discussions of the merits or lack thereof of various philosophical stances toward community investigation; and (2) technical criticisms of research, including much of my own work, guided by philosophical views that they find poorly suited to community ecology. Accordingly, my response will be in two corresponding sections.

PHILOSOPHY

Roughgarden believes that an investigator establishes a fact in science just as in everyday life, "by building a convincing case for the fact" (p. 583), through his native abilities, common sense, and experience. In everyday life, he feels, we rarely if ever adhere to formal rules in constructing a convincing case, so it is rarely if ever appropriate for scientists to abide by formal rules. I am sympathetic to Roughgarden's desire to ignore the entire corpus of epistemology and the philosophy of science. After all, so much formal writing by philosophers seems tedious, contrived, and supercilious. "Every philosophy . . . is liable to degenerate in such a way that its problems become practically indistinguishable from pseudoproblems, and its cant, accordingly, practically indistinguishable from meaningless babble" (Popper 1963, p. 7). Further, as Roughgarden implies, we seem to get along quite well in everyday life by using common sense and experience to construct a world view, so why not apply the same approach to our scientific endeavors and let tendentious philosophers argue about how many angels can dance on the head of a pin? If only for the time and mental effort saved, this is a tempting proposition.

Unfortunately, it is probably an overly simplistic one as well. For example, using common sense, experience, and native abilities, millions have built a convincing case for the existence of a deity, some of them proceeding very systematically to this conclusion (e.g., James 1902). Others have concluded by the same method that no such being exists, while agnostics have argued, equally commonsensically and (to themselves, at least) cogently, that there is no possibility of correctly reaching either of the other two positions. Are scientists doomed to such impasses, or are there formal procedures that promise a more satisfactory resolution? At least where experiment is possible, it seems to me that application of Popper's procedure (1963, 1972), clearly stated hypotheses and rigorous attempted falsification, is more likely to get us closer to an accurate account of nature, and to do so efficiently. Platt (1964) provides a nontechnical description of how this procedure, which he terms "strong inference," has spurred a remarkable series of successes in areas of physics and molecular biology and argues that it is the strict formalization that is responsible.

Now it is true that Popper's procedure could well be part of a commonsensical construction of a convincing case, so the two approaches need not be viewed as mutually exclusive alternatives. It is equally true that the search for confirmatory evidence is easier than the search for falsification and is very seductive. I suspect that in everyday experience it is by far a more common modus operandi than attempted refutation, and I feel that in community ecology it also holds sway. This is the detriment I see to rejecting formalization in favor of common sense. If one agrees that a procedure is effective and logically correct, its formalization is simply an aid to force us continually to match our actions and thoughts to the procedure. An analogy to probability theory is apt here. Most people feel that they have an intuitive or commonsense notion of the likelihood of different outcomes in simple but well-framed questions that arise in everyday life. A friendly game of blackjack is likely to be played, if not completely intuitively, at least with little more system than recalling the preceding few cards played, and with some attention to hunches. Yet we are very inclined to rely on formal statistical procedures when questions become more complex, even though the procedures themselves rely on the same principles that we ascribe to common sense. If we set out to make our living by beating the house consistently at blackjack, we would be much more likely to succeed if we rigorously employed a counting procedure and not the methods we use at home. In exactly this way, formalization of scientific procedures in community ecology need not be antithetic to common sense, but ought to help us to keep track of what we have and have not established in our study of what must be a vastly more complicated entity than molecular biologists or blackjack players deal with. Also, if one agrees with me that confirmatory evidence per se is not very compelling since one can always find some (Northrop 1948; Popper 1963), formalization ought to help us to be more efficient in allocating our research time.

It is ironic that Roughgarden should call for adherence to common sense as opposed to the Popperian procedures that my colleagues and I have advocated while a number of persons are contending that Popper's ideas have application well beyond the realms of natural science, including day-to-day activity. The art critic Gombrich (1960, 1973) and neurophysiologist Gregory (1973), for example, suggest that perception itself is a tacit construction of a hypothesis, with imagination and/or experience filling in those parts of a percept that have not been directly established by the senses. Popper himself (1972) noted this psychological variant of his refutation procedure. When a perceptual hypothesis is incorrect, it is an illusion that can mislead us in our everyday actions. Our sensations of the world constantly provide data that can be used to assess the hypothesis. To the extent that we ignore conflicting data and accept those data that tend to confirm our illusion, we are not thinking efficiently and our everyday actions are more likely to be misguided. There is nothing mysterious about this psychological model; it is simply meant to show that in everyday life as in science, "building a convincing case" by common sense and experience is not a trivial, automatically efficacious matter. And in everyday life as in science, application of a falsification procedure is likely to help.

Hintikka (1969, 1975) suggests a "possible worlds" model of epistemology, wherein any quest for knowledge, be it in everyday or scientific affairs, begins with a large number of depictions of reality (possible worlds), all consistent with "facts" that we agree are "established." Any advance consists in eliminating some possible worlds, thereby narrowing the scope of potentially valid depictions. Such a model, whose similarity to Popper's prescription for scientific research is apparent, seems eminently reasonable to me and indicates what constitutes an efficient procedure in science as in everyday life: one that invalidates a large number of hitherto possible worlds. It also suggests to me that a major pitfall in both everyday and scientific reasoning is failure to posit initially a sufficiently large universe of possible worlds, so that the real world is not among those initially present.

Particularly with respect to evolutionary community ecology theory, I feel that the Popperian prescription should be moderated, as Lakatos (1970) suggests, into a "sophisticated falsificationism." After all, the history of science is rife with observations that originally seemed anomalous but in the light of later developments were shown to be quite consistent with the hypotheses that spawned them. The complexity of communities and difficulty of manipulating them dictate that a theory not be immediately discarded because a single observation seems to refute it. As Lakatos (1970, p. 179) says, "one must treat budding programmes leniently," especially while they seem to be growing and generating research that anticipates new facts. But eventually there must be a willingness to confront theory with contradictory data in a decisive manner; Lakatos (1970, p. 179) allows only that "a budding research programme . . . should be sheltered *for a while* . . . " (my italics). My complaint is that community theory has been sheltered for long enough and has little more to show for this forbearance than a proliferation of theory. This is not a novel concern (cf. McFadyen 1975; Smith 1976; Futuyma 1975; Brown 1981; Pielou 1981), and I do not mean to be dogmatic about it. It may be, as Roughgarden feels, that we are on the threshold of dramatic, rigorous tests of community theory that will finally provide community ecology with well-tested, predictive maxims. If so, this past decade and a half of intense theoretical activity will have been well spent. If not, a large fraction of a generation of evolutionary ecologists will have spent its time manipulating models that are practically irrelevant to nature.

Having stated my reasons for desiring some sort of falsification procedure, even if delayed, I find much in Roughgarden's philosophical section with which I agree. Roughgarden is correct to distinguish between mathematical and biological proofs in that, exactly as Lakatos emphasizes, subsequent observations can always change our interpretation of a physical observation; but as Lakatos argues, this condition dictates only patience and leniency, not eternal immunity from attempted falsification. I also agree with Roughgarden (cf. Simberloff 1980) that the varied and complex nature of communities prevents a simple "litmus test" for the historic occurrence of coevolution or interspecific competition. Popper (1972, 1980) feels that his procedure applies to both singular and historic events, but surely falsification is vastly more difficult without replication. This difficulty in no way obviates the ultimate attempt to refute a theory, however, and it also suggests monumental modesty in assessing the scope and value of a model.

I even agree with Roughgarden and with Quinn and Dunham that we were incorrect to suggest (Strong et al. 1979) that the hypothesis of no population interactions has "logical primacy" as a null hypothesis. I do feel, however, that such a hypothesis is an apt starting point. There is such a plethora of information in the population biology literature on the ecological and evolutionary responses of individual species to habitat variation that it seems reasonable to assume that the habitat must always affect a population's density whether or not species interactions are important. In fact, as an appeal to Occam's razor (admittedly a psychologically attractive rather than a logically necessary principle) it seems appropriate to ask if habitat variation might not explain a species' distribution without any need to invoke other species' distributions at all.

If this were so, it would certainly not demonstrate that species interactions are not important, nor have my colleagues or I ever suggested that it would. But it should force us to recognize that our universe of "possible worlds" is larger than we had thought. If, on the other hand, we reject this null hypothesis (as my colleagues and I have often done), we have successfully narrowed this universe. By contrast, approaching this matter as a problem of everyday circumstances, as Roughgarden suggests, seems often to have led to no null hypothesis at all, in which instance we cannot in principle have narrowed the universe. Other times, "common sense" has led to an initial universe that excludes an eminently reasonable galaxy of possible worlds. Pielou (1981, p. 24) observes that theoreticians modeling multispecies systems routinely assume that competition is occurring and that such models ". . . shackle thought. With depressing frequency, they cause students to assert that this or that process must (or cannot) be taking place, merely because some model or other says it must (or cannot). So-called counter-intuitive models have the desirable effect of jolting thought out of the ruts created by earlier models . . . ."

I agree with Roughgarden, too, that it is a distraction to focus more on testing a simplifying model than on testing hypotheses about processes in an actual system. My question to him then is, How many of the pages on evolutionary ecology theory in, for example, *The American Naturalist*, are misguidedly focused on theory and how many are actually required to present the model that motivates tests of hypotheses? Roughgarden feels that the major culpability for theory's bad press rests on the "always naive" and "usually incorrect" set of "popular impressions about 'what theory actually does say' " (p. 598). I presume that by "popular impressions" he refers to interpretations by us benighted nonmathematicians and by the nonecologists who have to determine what practical lessons ecology yields. I would point out first that there is some blame attaching to the modelers themselves (cf. Levin 1975) and second, that the bloated theoretical literature has consequences beyond being a chore for ecologists to wade through.

Simply by persistence and volume alone, a field acquires both a life of its own and at least temporary respectability among nonexperts. That ecological modeling for its own sake is now a recognized discipline is witnessed by the emergence of journals (e.g., *Ecological Modeling, Theoretical Population Biology*) devoted primarily or solely to modeling and scarcely at all to whether the models correspond to nature. Now, I agree with Roughgarden that theoretical and empirical research need not march in lockstep; however, I would emphasize his qualifier: "Nonetheless, theoretical ecology is ultimately valued by its utility in understanding ecological processes in nature" (p. 597). I suspect that a good many ecological modeling attempts are not motivated by this ultimate goal. Pielou (1981) observes that many ecological models seem to have been constructed with no purpose in mind, but I believe she overstates this case. No biological purpose perhaps, but some purpose nonetheless. One would surely not go through the trouble of doing the research, writing the paper, and attending to the numerous tedious details that ensure its publication if one only wanted to fill up time. Many purposes in addition to biological ones may be served by doing such work: amour propre, professional prominence and its attendant rewards, even such mundane matters as tenure and promotion can all benefit from doing this work. Such goals are neither blameworthy nor unique to ecological modeling; Merton (1973) makes a strong case that they motivate all scientists. The important point is that such factors impart a tenacity to a discipline such that it withers slowly even when it turns out to yield few substantive results.

Alas, it is no easier for nonecologists to assess the validity of an ecological school than it is for ecologists to judge the relative merits of different approaches to cosmology or cancer etiology. So long as its practitioners have academic credentials and are sufficiently adamant and vocal, a patina of respectability accrues to a school in the eyes of nonspecialists, even scientists. I would argue that the enhanced stature of ecology in the eyes of other sciences that Roughgarden perceives (but does not document) is, if it exists at all, not informed approbation of ecology's accomplishments. More likely, it would be simply a comfortable feeling that some ecologists are speaking in the mathematical terms that physical scientists understand, rather than in Latin nomina. "Physics-envy" (Cohen 1971) is misguided; ecologists' proper goal should be not approbation from physical scientists but a firm understanding of natural processes, to the point where we can predict the outcome of specified ecological processes and answer many of the specific ecological questions of direct application that currently besiege us.

Respectability among nonspecialists has consequences beyond the enhancement of self-esteem, unfortunately. In the absence of a clear consensus among workers in the field, a recommendation supposedly dictated by "theory" can be promoted as policy when "theory" could as well be written or construed to generate exactly the opposite recommendation. For one example, the International Union for Conservation of Nature and Natural Resources (1980) in its overview statement about what measures we must immediately take to stem an imminent disastrous increase in extinction rate, states that refuge design criteria and management practices should accord with the equilibrium theory of island biogeography. This is in spite of the facts that (a) the theory itself is increasingly viewed as not a particularly faithful respresentation of most nature (Gilbert 1981); (b) the main recommendation advocated in the name of theory—single large refuges rather than clusters of small ones—is not a consequence of the theory (Simberloff and Abele 1976; Higgs 1981); and (c) most empirical data that bear on the matter do not support the recommendation (Simberloff and Abele 1982). Were this recommendation taken seriously, it could really hinder conservation efforts. I could easily cite other examples in such practical areas as pest and disease control, food production, etc., where ecological models have been used to support prominent recommendations that could as well be wrong as right. My point, however, is not to indicate weaknesses of specific theoretical recommendations. Rather, I want to suggest that exaggerated claims for ecological models can have consequences in the real world.

Finally, I must demur from Roughgarden's assertion that the Popperian philosophical stance is moot and irrelevant to the practice of scientific research. I can only hope that my above discussion of the benefits of systematization and rigor in reasoning about empirical facts, admittedly as much an assertion as Roughgarden's advocacy of common sense, is the more cogent.

Nor can I concur with Quinn and Dunham that, effective as Popper's procedure is in some sciences, it is inappropriate in ecology and evolution because the latter fields are characterized by multiple causality. I agree that any natural historic phenomenon is very likely to have been affected by a number of forces, but I suspect that this complication only demands more ingenuity on the part of researchers in framing unambiguous hypotheses. After all, molecular biological processes are also affected by a variety of inputs, yet Quinn and Dunham do not contest Platt's estimation (1964) of the success of strong inference in that field.

Many of the crowning achievements of molecular biology were achieved in spite of complicated inputs, by clever experimental procedures that could unambiguously test well-chosen hypotheses. In genetics it is routine to use mutants that cannot manufacture a particular molecule in order to show an effect that would otherwise be masked. Avery et al. (1944) demonstrated that DNA was the active agent in bacterial transformation by using a *Pneumococcus* strain that does not manufacture a polysaccharide coat. Benzer (1962) used mutants of T4 phage, the rII types, that could not grow on strain K of *Escherichia coli* to demonstrate many aspects of genetic structure. Many other definitive experiments in physiology and molecular genetics rested on inhibiting or repressing agents that affected one or more molecular functions while leaving others intact. Analogous techniques should allow ecologists to test hypotheses in similarly incisive fashion. For example, at a stroke, Keith (1963) was able to refute the hypothesis that snowshoe hare cycles are part of an intrinsic predator-prey oscillation when he found that hare populations oscillate on Anticosti Island, in the absence of lynx. Underwood (1978) caged three grazing gastropod species in a variety of densities to show clearly competitive effects of *Cellana tramoserica* on *Bembicium nanum*, and equally clearly the absence of other competitive effects among the species. For

Underwood's result as for Keith's the key to a clear result was an unequivocally falsifiable hypothesis and a system sufficiently simplified, by whatever means, to allow an unambiguous test of the hypothesis.

It is not necessary, for a hypothesis to be falsifiable, that it state that force A and only force A is acting, while the alternative hypothesis posits only force B. Rather, it is at least as much a matter of stating the hypotheses and choosing the tests in such a way that their outcome is unambiguous. It does not make sense (nor have my colleagues or I attempted this) to hypothesize that competition and only competition is responsible for a community pattern, and to pose as alternatives that predation and only predation, weather and only weather, etc., are uniquely responsible. It may be a very difficult task in community ecology to state informative hypotheses that allow unambiguous answers, but it is probably worth the effort (Strong 1980). If the bone of contention is some claimed historic or very infrequently occurring phenomenon the resolution of the matter by Popperian methods will be even harder, but I doubt that the inductive or descriptive method advocated by Quinn and Dunham—estimating the probable contributions of a number of a priori "causes"—will provide an easier path. Such a method seems to me to resemble Roughgarden's commonsense approach, and would thus suffer the same shortcomings. As I said above, I could well be wrong, and only the collective judgment of ecologists over the next few years will determine if either variant produces convincing and useful ecological descriptions.

I cannot refrain from protesting about two points of fact. First, nowhere did we (or Popper or Lakatos, for that matter) argue that scientific progress will occur faster if one investigates alternative hypotheses rather than the principal thesis of interest. This view of procedure as an either/or phenomenon erects a false dichotomy. Our contention has been that there must be more than one hypothesis (null and alternatives, if you will) such that the empirical data sought will argue for one over the others: the very procedure described by Platt (1964) for molecular biology. Second, Quinn and Dunham state, "Popper (1959, but see 1983a, 1983b) argued explicitly that estimates of probability are unfalsifiable, and thus not subject to scientific test under criteria of disproof" (1983, p. 603). Popper (1972, a revision of 1959) states: "I believe that my emphasis upon the irrefutability of probabilistic hypotheses . . . was healthy: it laid bare a problem which had not been discussed previously . . . . Yet my reform . . . changes the situation entirely. For this reform . . . amounts to the adoption of a methodological rule . . . which makes probability hypotheses falsifiable" (p. 191).

TECHNICAL CRITIQUE

As sympathetic as I am to Roughgarden's desire to obviate the philosophy of science, I cannot agree with much of his technical critique. A fair fraction of it seems to me to exemplify the shortcomings of a commonsensical approach. For example, Roughgarden's outline of coevolutionary theory appears (by virtue of niches' being to the right or left of one another) to assume a one-dimensional niche space, surely an unwarranted restriction on the number of possible worlds. That one-dimensional results need not trivially and automatically translate into multidimensional ones is well known (e.g., May 1974). And nowhere in his criticism of Connell's protocol does Roughgarden suggest any conceivable observation that would be taken to indicate that exploitation competition was not important in a particular system. Then why do any field or laboratory research at all? It is certainly true that the removal experiment advocated by Connell (1980) cannot in principle exorcise the ghost of competition past, but it seems as if no data can, even though we are not dealing here with a logical proof. At least Connell's procedure has the virtue of clarifying how nature presently operates.

Roughgarden again seems to narrow the universe of possible worlds unwarrantedly when he criticizes our null hypotheses because "No biological processes are exhibited that produce the distributions predicted by [them]" (p. 591). Although we do not attempt to account for every force acting on a species, nor for the exact mechanism by which forces act, it is clear that we are asking if species' individual responses to the physical environment suffice to explain their distributions. Roughgarden similarly omits likely important forces from his coevolutionary model. (Is he saying, e.g., that weather is unimportant, that the habitat is homogeneous?) He is also not very specific about mechanisms (e.g., is the competition exploitation or interference?). Generally he seems to be arguing that a model, to be useful, must be realistic; but he has not noticed that all models are unrealistic by virtue of being abstractions (Simberloff 1984); that the dividing line between "realistic" and "unrealistic" is subjective (Pielou 1981); and that unrealistic models are at least as useful as realistic ones in advancing our understanding of nature or generating mechanistic hypotheses (Pielou 1981; Simberloff 1984).

In testing null and alternative hypotheses based on more or less unrealistic models, what is important is that different degrees of realism not affect the models' performances in the area of concern. Roughgarden subsequently seems to accept this view when he avers that most community ecology models aim to simplify, and that "we fill the gaps in the simplified description with assumptions that may actually be false if taken literally, but which are hoped somehow not to be too misleading" (p. 594). I do not see that Roughgarden's coevolutionary models depict the "real processes that bring species to islands" any more than ours do, yet he feels we must model these forces to determine what distributions of data are expected in the absence of interspecific competition. Why must he not also model them to determine what distributions are expected in the presence of interspecific competition?

In no sense have my colleagues and I ever said the world is without structure. Nor have we intended that a pattern must be manifest before one is licensed to pursue research into what may have caused the pattern. Our sole concern has been that the full range of forces potentially generating the pattern be recognized so that the search for its causes not be unduly circumscribed initially. Certainly a process may be studied fruitfully before its consequences are well understood, so long as the process does not imperceptibly become established as fact simply by virtue of its study or its mathematical tidiness. Roughgarden does not recognize this as a problem: "Sometimes it is obvious that a process is occurring" (p. 592). No doubt as obvious as is the existence or nonexistence of God.

The "serious technical" criticisms of our work to which Roughgarden alludes are fully discussed by Strong and Simberloff (1981), Simberloff (1984), and Connor and Simberloff (1984), who have shown them to be either incorrect or of no consequence with respect to our conclusions. Nor has Roughgarden statistically substantiated his claim that our tests are biased toward type II errors. I have yet to see his analysis of bias or power of any of his models or ours. Related complaints by Quinn and Dunham are similarly dismissed in the above rebuttals. In particular, the observation that a source pool may already be structured by competition was anticipated by Connor and Simberloff (1979) and Strong et al. (1979), and is not germane to the hypotheses they were testing, nor does it generate circularity. Its sole potentially debilitating effect is to lower statistical power. With respect to presence and absence of species combinations on islands, Connor and Simberloff (1979) were explicit that the occurrence frequencies of the different species might well be partly determined by interactions, but that the question asked was if the co-occurrence distribution patterns of the different species, above and beyond the individual frequencies, demanded an interactive explanation: ". . . *unless* one is willing to ascribe to competition the facts that islands have different numbers of species and that species are found on different numbers of islands, the New Hebrides data still argue heavily against the claim that competition determines most aspects of the distribution of species on islands" (p. 1,136, my italics). As for the contention that the bill sizes in a species pool may already be structured by competition, Strong et al. (1979) did not dispute this. All they did was to perform the classical statistical procedure of randomization (Bradley 1968; Hendrickson 1981) to see if observed subsets of birds on islands were typically different in bill-size characteristics from those we would have observed if species and races were independently and randomly placed on islands. Again, we did not contend that sizes of birds in the species pool were not affected by competition; we asked only if the sizes of whichever races coexist, above and beyond the distribution of sizes in the pool, additionally demand an interactive explanation.

LITERATURE CITED

Avery, O. T., C. M. MacLeod, and M. McCarty. 1944. Studies on the chemical nature of the substance inducing transformation of pneumococcal types. Induction of transformation by a desoxyribonucleic acid fraction isolated from *Pneumococcus* type III. J. Exp. Med. 79:137–158.
Benzer, S. 1962. The fine structure of the gene. Sci. Am. 206:70–84.
Bradley, J. V. 1968. Distribution-free statistical tests. Prentice-Hall, Englewood Cliffs, N.J.
Brown, J. H. 1981. Two decades of homage to Santa Rosalia: toward a general theory of diversity. Am. Zool. 21:877–888.
Cohen, J. E. 1971. Mathematics as metaphor. Science 172:674–675.
Connell, J. H. 1980. Diversity and coevolution of competitors, or the ghost of competition past. Oikos 35:131–138.
Connor, E. F., and D. Simberloff. 1979. The assembly of species communities: chance or competition? Ecology 60:1132–1140.
———. 1984. Neutral models of species' co-occurrence patterns. *In* D. R. Strong, Jr., D. Simberloff, L. G. Abele, and A. B. Thistle, eds. Ecological communities: conceptual issues and the evidence, Princeton University Press, Princeton, N.J. (in press).
Futuyma, D. J. 1975. Review of M. L. Cody's Competition and the structure of bird communities. Q. Rev. Biol. 50:217.

Gilbert, F. S. 1980. The equilibrium theory of island biogeography: fact or fiction? J. Biogeogr. 7:209–235.
Gombrich, E. H. 1960. Art and illusion: a study in the psychology of pictorial representation. Pantheon, New York.
———. 1973. Illusion and art. Pages 193–243 *in* R. L. Gregory and E. H. Gombrich, eds. Illusion in nature and art. Scribner's, New York.
Gregory, R. L. 1973. The confounded eye. Pages 49–95 *in* R. L. Gregory and E. H. Gombrich, eds. Illusion in nature and art. Scribner's, New York.
Hendrickson, J. A. 1981. Community-wide character displacement reexamined. Evolution 35:794–809.
Higgs, A. J. 1981. Island biogeography theory and nature reserve design. J. Biogeogr. 8:117–124.
Hintikka, J. 1969. Models for modalities. D. Reidel, Dordrecht, The Netherlands.
———. 1975. The intentions of intentionality. D. Reidel, Dordrecht, The Netherlands.
International Union for Conservation of Nature and Natural Resources. 1980. World Conservation Strategy. IUCN-UNEP-WWF.
James, W. 1902. The varieties of religious experience. Longman's, Green, New York.
Keith, L. B. 1963. Wildlife's ten-year cycle. University of Wisconsin Press, Madison.
Lakatos, I. 1970. Falsification and the methodology of scientific research programs. Pages 91–196 *in* I. Lakatos and A. Musgrave, eds. Criticism and the growth of knowledge. Cambridge University Press, Cambridge.
Levin, S. A. 1975. On the care and use of mathematical models. Am. Nat. 109:785–786.
MacFadyen, A. 1975. Some thoughts on the behaviour of ecologists. J. Anim. Ecol. 44:351–363.
May, R. M. 1974. Stability and complexity in model ecosystems. 2d ed. Princeton University Press, Princeton, N.J.
Merton, R. K. 1973. The sociology of science. Theoretical and empirical investigations. University of Chicago Press, Chicago.
Northrop, F. S. C. 1948. The logic of the sciences and the humanities. Macmillan, New York.
Pielou, E. C. 1981. The usefulness of ecological models: a stock-taking. Q. Rev. Biol. 56:17–31.
Platt, J. R. 1964. Strong inference. Science 146:347–353.
Popper, K. R. 1959. The logic of scientific discovery. Hutchinson, London.
———. 1963. Conjectures and refutations: the growth of scientific knowledge. Harper & Row, New York.
———. 1972. The logic of scientific discovery. 3d ed. Hutchinson, London.
———. 1980. Letter to editor. New Sci. 87:611.
Quinn, J. F., and A. E. Dunham. 1983. On hypothesis testing in ecology and evolution. Am. Nat. 122:602–617.
Roughgarden, J. 1983. Competition and theory in community ecology. Am. Nat. 122:583–601.
Simberloff, D. 1980. A succession of paradigms in ecology: essentialism to materialism and probabilism. Synthèse 43:3–39.
———. 1984. Properties of coexisting bird species in two archipelagoes. *In* D. R. Strong, Jr., D. Simberloff, L. G. Abele, and A. B. Thistle, eds. Ecological communities: conceptual issues and the evidence. Princeton University Press, Princeton, N.J. (in press).
Simberloff, D. S., and L. G. Abele. 1976. Island biogeography theory and conservation practice. Science 191:285–286.
———. 1982. Refuge design and island biogeographic theory: effects of fragmentation. Am. Nat. 120:41–50.
Smith, F. E. 1976. Ecology: progress and self-criticism. Science 192:546.
Strong, D. R., Jr. 1980. Null hypotheses in ecology. Synthèse 43:271–285.
Strong, D. R., Jr., and D. Simberloff. 1981. Straining at gnats and swallowing ratios: character displacement. Evolution 35:810–812.
Strong, D. R., Jr., L. A. Szyska, and D. S. Simberloff. 1979. Tests of community-wide character displacement against null hypotheses. Evolution 33:897–913.
Underwood, A. J. 1978. An experimental evaluation of competition between three species of intertidal prosobranch gastropods. Oecologia 33:185–202.

NATURAL VARIABILITY AND THE MANIFOLD MECHANISMS OF ECOLOGICAL COMMUNITIES

Donald R. Strong, Jr.

Department of Biological Science, Florida State University, Tallahassee, Florida 32306

Submitted November 5, 1982; Accepted April 14, 1983

PART 1

In response to Roughgarden's provocative essay, the editor of *The American Naturalist* has solicited articles that present the personal perspectives of some community ecologists and their sets of operating assumptions. Ecology is growing and changing rapidly, as is shown by this dispute over fundamentals, and my concern will be the importance of alternative and complementary forces in communities, in contrast to a singular emphasis on competition.

I will emphasize density-vague population dynamics; independent and noncompetitive coexistence of species; weak and inconsequential competition; variable and diffuse species interactions; maintenance (by the weather, disturbance, and natural enemies) of populations well below densities that deplete resources to any important extent; and mutualistic and commensalistic relations between species. My purpose is to illustrate the commonness of communities that are quite different from those assumed by orthodox competition theory.

I certainly do not deny that competition occurs in nature or that it is in some cases a dominant process. Competitively dominated communities have been amply reviewed elsewhere. Strong and persistent competition, however, does not necessarily square a set of species with orthodox competition and niche theory. Effects of competition are varied, and mathematical deductions about them need critical testing, both empirical and theoretical. Though interspecific competition, when and where it occurs, may affect aspects of population dynamics, shape niches, cause coevolution, or even result in character displacement of some species in some circumstances, it need not have any of these effects. What are the possibilities for diffuse competition (Connell 1975)? What might nonequilibrium communities be like (Caswell 1982; Wiens and Rotenberry 1980)? What are the influences of habitat patchiness and high stochasticity (Levin 1976; Chesson 1978; Murdoch 1979)?

My paper is in two parts. Part 1 deals with several logical and philosophical points. Part 2 is a description of some case studies that find factors other than interspecific competition to be of great influence in ecological communities. The dispute illustrated by this symposium was actually begun long ago (British Ecological Society 1944). This dispute, I believe, can be resolved only by critical empiricism and by recognition of the variable nature of ecological communities. Other reviews with a perspective similar to mine are those of Connell (1975), Shapiro (1975), Birch (1979), Simberloff (1981), Wiens (1983), and James (1982).

ECOLOGY NEEDS MORE THAN COMMON SENSE

The exhortation to use common sense (Roughgarden 1983) for establishing facts in science is curious. "Common sense is that which tells you that the world is flat" (Chase 1938). Yes, common sense sometimes leads to sound judgment, but it is also ordinary, free from intellectual subtlety, not dependent on special or technical knowledge, and it is unreflective opinion (Lewis 1961). We have *Common Sense About Yoga* (Pavitrananda 1944), *The Common Sense of Drinking* (Peabody 1931), and *Common-Sense Suicide* (Portwood 1978). A long dispute in philosophy pits common sense against unorthodox and progressive ideas (Berkeley 1713; Lewis 1961).

Common sense is often synonymous with conventional wisdom, which can be quite useful and give correct judgments, especially about common, ordinary, and conventional things, things with which many people have had a great deal of empirical experience. On the other hand, unusual and poorly understood things, and things with which people have little direct experience, are often not well understood by means of ordinary, day-to-day responses. Just as very small and immense phenomena in physics are unusual and need uncommon means of analysis, the very subtle, diffuse, changeable and complex phenomena in ecological communities are not well understood by just ordinary means.

An approach that uses more than common sense and orthodox thought is necessary in ecology because much of our science deals with phenomena with which few people have the opportunity to develop much practical experience. It is true that foresters, hunters, fishermen, and other people who spend much time outdoors can develop deep understanding of facets of nature, but even these people have little access to much knowledge, thinking, and tools useful for understanding ecological processes. For example, the ecosystem processes of energy flow and nutrient cycling were not uncovered by common sense, but rather by thinking and experiments that were quite uncommon in the context of their time (e.g., Lindeman 1942; Likens et al. 1977). Good experimental demonstrations of competition operating in nature (e.g., Connell 1961) and of how predation (Paine 1966) and abiotic factors (Dayton 1971; Tinkle 1982) can greatly modify and lessen influences of any competition in communities were based upon perceptions and experiments so uncommon that they generated tremendous insight for ecology.

Common sense is silent about two elements of science that I consider to be essential to ecology: imagination and testing. In fact, common sense has often been hostile to scientific imagination, for example, when it opposed the heliocentric system of planets, the circulation of blood, the great age of the earth, and evolution.

CURIOSITY, TESTING, AND VERACITY

Boulding (1980) gives a broad and tolerant definition of science as curiosity, testing, and veracity. He emphasizes that science increases knowledge "not by the matching of images with the real world (which Hume pointed out is impossible) . . . but by a relentless bias toward the perception of error" (p. 832). To me, science is much more than common sense building a case for a fact, as Roughgarden says. "A widespread illusion about science is that its basic theoretical images and paradigms are the result of inductive reasoning and experiments" (Boulding 1980, p. 830). To me, the essence of science is captured in one last quote from Boulding: "It would be truer to say that science is the product of organized fantasy about the real world, tested constantly by an internal logic of necessity and an external record of expectations, both realized and disappointed" (p. 833). Of course this statement means that science is practiced by people other than "scientists." It is practiced in varying degrees by farmers, investors, athletes and teams, fishermen, and others.

My and many others' rule of thumb about testing is to ask the question, "What material evidence would weigh against this idea?" If I cannot come up with any, I do not necessarily discard the idea, but try to rephrase it in a way that makes it possible to imagine measurements, experiments, or comparisons that could yield results that would contradict the idea. As Popper (e.g., 1962) and others have pointed out, the tougher the test, the surer we can be about an idea that passes a test. This concept can be expressed as riskiness of predictions; risky predictions that are confirmed increase our regard for a theory much more than do safe predictions, which are likely to be corroborated for reasons quite independent of the details of the theory. Our regard should be greatest for theories that have passed multiple, independent, tough tests. Much of my concern about ecological theory in general, and competition theory in particular, is that their predictions are not usually risky. Vigorous defense of theories that do not yield risky predictions seems to me to be a waste of scientific time and effort. If the predictions of theories are vague, the science is weak.

NULL HYPOTHESES

I am sympathetic to Quinn and Dunham's (1983) point of view; ecological null hypotheses are often quite difficult to state because of the complexity, history, and multiplicity of ecological factors (Strong 1980). Sciences can be compared along an axis that ranges from atomistic to organismal, with ecology at or near the organismal extreme. The exemplars of science, such as physics, chemistry, and now molecular biology, are near the atomistic end, where universal processes, which usually dominate particular systems, have been discovered. The metaphor "atomistic" implies systems like atoms as we know them today, which, within categories, are simple, consistent, and composed of discrete parts with definite and distinct interactions. (I do not mean "atomistic" to imply irreducibility.) Ecology concerns organismal systems, which usually are as different as can be from atomistic systems. The components of ecological systems (species, populations, and individuals) have a multiplicity of mostly continuously varying characteristics rather than the many fewer and often discrete features of particles, atoms, or molecules. The most similar sets of ecological components are usually quite heterogeneous in many ways, while fundamental units in atomistic sciences are virtually uniform. Finally, history is a large part of phenomena in organismal sciences, while the influence of history is quite restricted in atomistic sciences; a carbon atom is a carbon atom, regardless of its previous molecular or crystalline associations (with minor variations in radioisotopes), but a population has at least age structure, genetic heterogeneity, and developmentally induced differences among individuals that are the product of history both recent and old.

These differences mean that phenomena occur in quite different contexts among sciences. In atomistic systems, processes are often obvious and contrasted against a background of sameness; in organismal systems processes often manifest themselves in subtle and ambiguous ways and occur along with other processes that produce similar results. Null hypotheses are particularly useful in ecology because simultaneous but different processes give signals so similar that the products of individual phenomena are difficult to distinguish. For these reasons, I believe that null hypotheses have logical primacy (Strong et al. 1979) especially in ecology.

The ambiguity and complexity of ecological processes that Quinn and Dunham describe is most effectively dealt with by statements of null hypotheses; what pattern, outcome, or change we would expect if a process were not operating, or if it were operating in a different way or at a quite different level. If one cannot conceive of an unambiguous null hypothesis, then one cannot critically assert from the system or data at hand that the process of interest is causing the effect. If a null hypothesis showing what a community of species would look or behave like in the absence of competition cannot be constructed, one cannot use the system as evidence for competition.

Operationally, I am very supportive of the attempts to frame null hypotheses, albeit complex ones, by Dunham (1980) and Dunham et al. (1978) and agree with these authors to the fullest that often the problem is discovering what effect competition has, or whether it is intense enough to have any influence at all, rather than just whether it has occurred.

AUTECOLOGY AND INDIVIDUALISTIC RESPONSES OF SPECIES

Ecological communities are groups of species living closely enough together for the potential of local interaction. Each of the species in a community has at least a slightly different geographical range and habitat and usually pronounced differences. The local ecology of each species is primarily set by its autecology, by species above and below in the food web, and by its migration. Abiotic forces of climate and weather, food and natural enemies, and movement of individuals into and out of the locale are the basic determinants of local species existence. Secondarily, interspecific competition may modify local species existence strongly, weakly, or not at all (Krebs 1978). This perspective is akin to the individualistic view of species in communities (Gleason 1926; McIntosh 1975).

Individualistic theories emphasize the primary importance of autecology, migration, and food for heterotrophic organisms in determining local existence of species in communities. Individualistic responses are most often cast as an alternative to the superorganism theory, which posits species as tightly knit, integrated components that accommodate each other as do cells or tissues in an organism. Individualism also is an alternative to another extreme, to excessive competitionism, which posits these negative interactions between species as the primary and overriding factor in communities.

Extreme competitionism makes an error of excessive emphasis that is in some ways the opposite of that of superorganism theories; communities are no more primarily the product of competition than they are the product of superorganismal integration. One of the goals of community ecology is to weigh critically whether competition, or other synecological factors, actually does modify individualistic species existences in particular cases, and if so, how much (Wiens 1983). Individualism gives a reference point, a way of framing a null hypothesis, for studying the effects of competition and other interactions between species. Until autecological facets of existence are understood, it is tenuous to infer much about synecological influences.

SPECIES INDIVIDUALISM, COMMUNITY "STOCHASTICITY," AND PATTERNS

Species individualism, independent coexistence of species as a function of different autecology and separate parts of food webs, has been termed "stochasticity" by Grossman et al. (1982) in a fascinating analysis of temporal population patterns of Indiana stream fish. They found no consistent patterns in ranks of species and no consistency in ranks of trophic groups over time by means of concordance analysis. After pointing out the great problems of fuzzy definitions for ecological terms, they propose "stochastic" to describe species assemblages that show no tendency of persistence or resilience, those in which species change greatly through time in presence-absence and relative abundance. In the context of the MacArthurian tradition of studying "patterns" as a means of inferring simple explanations of interspecific processes (MacArthur 1971, p. 190), "stochasticity" seems an appropriate term.

Stochasticity implies that at least one variable under consideration has a variance greater than zero; for real ecological systems, variances are often large (Simberloff 1980). Chesson (1978) gives a useful general discussion of ecological stochasticity. It is important for ecologists not to infuse unnecessary metaphysical problems into discussions about population and community influences. Stochasticity does not imply a lack of causation, and the rank of species and trophic groups could have little statistical concordance but still have deterministic causes. For stream fish, Schlosser (1982) presents a context in which determinism of several sorts probably would cause dynamics with little concordance. A lack of interspecific concordance does not necessarily imply that species are not greatly affected by interspecific interactions. For example, certain forms of mathematically deterministic population regulation can cause densities to cycle or even behave chaotically (May 1981) (although it is unclear whether "chaos" is an ecologically realistic expectation, especially for species with overlapping generations). Certainly, time lags in effects of species interactions could generate statistical discordance, even in communities strongly affected by competition between species. As well, species that were competing intensely, but with nonlinear interactions and fluctuating population sizes (Armstrong and McGehee 1980) might show little concordance.

One of the most distinct patterns of high concordance in the ecological literature occurs among species of oceanic copepods in the North Pacific (McGowan and Walker 1979). The authors' interpretation of this high concordance is: "Aspects of modern community theory, based upon competitive equilibrium, seem inadequate to explain our results. Predator regulation of structure seems a more likely explanation. However, existing information indicates a lack of sufficient specialization of what we believe to be the main predators on copepods to account for the observed constancy . . ." (p. 195). Thus, the cause of high concordance in this case is really unknown. One might argue that high concordance would be caused by forces quite different from those proposed by Grossman et al. (1982), that high concordance could be expected for independently coexisting species that were all greatly affected by the same overwhelming autecological factors. A series of dry years in a rainforest might cause a general decline in populations of species adapted to the normally wet climate, independently of any species interactions. A series of warm years in the arctic, or of cold winters in south Florida, would be expected to do the same.

The meaning of general patterns is a historically important approach to community ecology, but an approach that I fear now bears relatively little promise for understandng mechanisms. My reasons for this judgment are given above in the section on null hypotheses; quite different processes commonly produce what we perceive to be the same general pattern, be the pattern differences in species sizes or a lack of concordance in population dynamics. Conversely, mechanisms of competition may not produce any specific community effect. I agree with Poole (1977) that it is usually impossible to discern deterministic mechanisms from passive observations of population dynamics that have much of a stochastic component; the degree of concordance among species abundances need not reflect the degree of interspecific regulation. I believe that the role of population dynamics in community ecology cannot really be known without experiments, because population change can be caused by such a large number of factors.

DOES NICHE THEORY GUILD THE ECOLOGICAL LILY?

To gild refined gold, to paint the lily, . . .
[Shakespeare (*King John*, Act IV, iii. 11)].

In the extreme, niche theory can be typology, only interpreting nature in ways that are consistent with the theory and treating natural variation as bothersome noise rather than the signal of other important but inconvenient variables. In this sense, niche theory can be anti-Darwinian, for Darwin's operating principles were antitypological (Ghiselin 1969). In Peters' (1976, 1980) view, we have a "niche concept" but not much of a niche "theory," because these notions do not predict ecological phenomena in any but the weakest, softest, and most trivial fashion. After Rigler (1975), "concepts" are not predictive and are not falsifiable, at least in the eyes of their beholders. We do know that the mathematical inferences from niche theory are quite dependent on assumptions; rather than being general, niche theory is quite specific (Abrams 1976). As well, quantitative assessments of niche overlap can lack simple and consistent biological meaning and usually ignore variability in resource states (Hurlbert 1978). That natural variability and other biotic factors such as predation (Schoener and Schoener 1982) can relegate competition to a rather minor role in lizard communities is shown by Tinkle (1982) and others, discussed above.

Few studies have explicitly cast the theory of competitively constrained niches as an alternative hypothesis. Several alternatives to niche theory are conceivable, including communities with intense competition that is not played out on resource continua, as is the case of some sessile organisms (Buss and Jackson 1979; Buss 1981), especially those that are greatly affected by disturbance (Dayton 1971; Hartshorn 1980). An important alternative to niche theory applies to organisms that specialize on a distinct resource, but that do not attain sufficient densities to deplete resources significantly. This alternative is highlighted by many phytophagous insects, discussed below.

Aho et al. (1981) have viewed the theory of specialized niches constrained by competition as an alternative to an individualistic hypothesis, for snails of Finnish lakes. They found a view based on independent autecological adaptation among species more consistent with the data than one based on contiguous, competitively defined niches. This is not to say that competition does not occur, but that it is at least secondary to autecological adaptation to abiotic features of the environment in determining community composition.

MATHEMATICAL ECOLOGY: IS THE MEDIUM THE MESSAGE?

"Theoretical work often diverged too far from life and became exercises in mathematics inspired by biology rather than an analysis of living systems" (Levins 1968, pp. 3–4). In a similar vein, May et al. (1979, p. 268) describe their mathematical efforts as a "crude caricature of multispecies systems," and Oster (1975, p. 35) says, "I certainly would be the last to argue that mathematics is 'no good'; I make my living from it! I only mean to counsel caution in its application and a healthy cynicism about its efficacy in dealing with complex systems."

The literature appraising mathematics' role in ecology is vast and has a long history (McIntosh 1980), and several times in the past extravagant claims for mathematical ecology have been made. To me, mathematics plays a mildly interactive and mutualistic role with empirical ecology. Levin's (1981) model of this interaction, which stresses parallel and reactive mutualism between applied mathematics and ecology, describes a desirable relationship between the two disciplines.

Ecology would not be healthy without both vigorous empiricism and vigorous theory, but equally important is the interaction between the two. Without theoretical perspective, empiricism is doomed to do no more than chronicle idiosyncrasy, and insouciance toward the real world leads theory to irrelevance. As stressed by Levin (1981), the two components should be tolerant of one another; the course of theory cannot be dictated by empiricism, and vice versa. But these two sides of our family should try hard to influence one another, and the two should be intensely interested in each other. Efforts of one should be appreciated enough by the other that course changes and corrections can occur efficiently. Empiricists are becoming increasingly sensitive to mathematical theory's obsession with "generality" and obliviousness to the details of autecology: ". . . stability analyses appear to reflect the formulation, assumptions and parameter values of models rather than testable field biology" (Wang and Gutierrez 1980, p. 435). A good example of careful attention to the relationship between empiricism and theory is given by Getz and Gutierrez (1982), who are agriculturalists and have responsibility for the reliability of their theory; they need theory that is predictive.

CHARACTER DISPLACEMENT: A SHROUD OF TURIN FOR ORTHODOX COMPETITION THEORY

Ecological character displacement and coevolution are perhaps the most extreme extensions of orthodox competition theory. Like the Shroud of Turin for some of the most zealous Christians, the validity of character displacement and coevolution have long been too cherished for critical testing by some competitionists. Recent research, both empirical and theoretical, however, creates substantial doubt that these are general phenomena in nature. Connell (1983), in this volume, addresses literature and logic on this subject. Theoretically, there are substantial reasons for skepticism of character displacement and many claims of coevolution, on both ecological and evolutionary grounds. Empirically, no solid evidence has emerged in favor of character displacement, and several tests with particular faunas have tended to rule it out.

On the basis of genetical theory two authors have recently proposed that the evolution of character displacement may be rather unlikely (Slatkin 1980; Arthur 1981). On the basis of ecological theory, two lines of modern evidence weigh heavily against the general occurrence of character displacement. First, differences in animal morphology do not usually correlate as finely and consistently with differences in food as would be necessary for general character displacement among species (Willson 1971; Wilson 1975; Hespenheide 1971, 1973, 1975). Huey (1979) carefully details the problems of interpreting character differences as character displacement for some South American geckos.

One of the most complete tests of the assumption that interspecific competition is a simple function of similarity in body size was done by Dunham et al. (1978), with lizards in the genus *Uta* from islands in the Gulf of California. Like allied work by Tinkle and his students, this study is based upon a wealth of experience with these and similar species. Their null hypotheses were built around noncompetitive explanations for adaptive variation in body size, around individualistic, independent species reactions to food differences, intraspecific competitive differences, sexual selection, and food web factors such as predation. "The results not only supported the null hypothesis of no competitive effect, but demonstrated that independent variables not obviously associated with competition explained at least as much of the variance in *Uta* body size as did variables directly related to competition" (Dunham et al. 1978, p. 1,230).

Hairston (1981) experimentally showed how a priori deductions based on taxonomic or morphological similarity are inadequate indications of interspecific interaction for some salamanders. Species removals over a 5-yr period produced increased performance in some but not in other closely related species. "The results call into question the common assumption that competition is the organizing force in all coexisting assemblages of species which share common resources" (p. 65). When competition occurs among these temperate-zone salamanders, it seems to me that interference (rather than exploitation and depletion of food resources) is most likely the mechanism (Jaeger 1974; Hairston 1981; Keen 1982), but others believe that the "jury is still out" (A. Dunham, personal communication).

A stunning recent finding concerns Galapagos finches (Schluter 1982), which have long been considered to be exemplars of character displacement. On Isla Pinta, *Geospiza fuliginosa* and *G. difficilis* are the most similar pair in beak depth and culmen length. Lack (1947) inferred on this basis that they would be the strongest competitors for food. Rather than using the most similar food and thus having the strongest potential for interspecific competition, these two species use quite dissimilar food; *G. fuliginosa* is largely granivorous, and *G. difficilis* is largely carnivorous, eating arthropods and gastropods that live in leaf litter. Schluter (1982) learned that the birds behave in a manner quite consistent with Gleason's individualistic hypothesis. The different species move rather independently up and down the island in response to changes in their food supply. Although some overlap in food does occur, it does not have great interspecific effects. "Factors other than competition, mainly variations in the food supply itself, are far more important in determining altitudinal distributions and other attributes of finch populations" (Schluter 1982, p. 1,515).

Other recent studies reinforce the idea that interspecific competition is not usually a function of fine morphological differences between closely related species. For a pair of surfperch species in the genus *Embiotoca*, diet differences are much better indicated by differences in foraging behavior than by differences in morphology (Schmitt and Coyer 1982). For sparrows, a decade-long quest that critically weighed various models, found that differences in behavior, resource fluctuations caused by factors other than the birds themselves, and apparently idiosyncratic autecological differences among species explain interspecific influences much better than models based upon size differences between species (Pulliam 1983).

A second major line of evidence that autecology and individualistic responses of species, independent of potential competitors, are the main sources of morphological differences comes from studies with climate and ecogeographic variation, from gradients in size, shape, and color of species (Bergman's rule, Allen's rule, Gloger's rule; James 1970). "Variation in morphology across a species range is largely a function of climatic conditions" (Fleischer and Johnston 1982, p. 747).

Sexual selection is also a plausible autecological explanation of intraspecific differences in morphology (Downhower 1976). After geographical variation that is presumably autecological in origin is factored out, several of the supposed cases of ecological character displacement no longer appear valid (Grant 1972).

Short-term changes in climate and weather produce quick changes in mean population morphology and greatly reinforce our confidence that selection by autecological factors accounts for a substantial fraction of intraspecific variation in morphology for Galapagos finches (Boag and Grant 1981) and for house sparrows (Fleischer and Johnston 1982, p. 747). Certainly, if great stochastic changes in morphology are consistently thrust upon populations as a result of climatic fluctuations, the fine tuning of morphology that is supposed to derive from interspecific competition is much less likely. Selection for character displacement would have to be quite intense to be felt through the strong but stochastic forces of selection upon morphology imposed by climatic fluctuation. Theoretically, the blade of a pocket knife could steer a large vessel over an ideally placid sea, but the tossings and cross currents of wind and waves require much larger rudders on vessels in real seas.

Experimental work with character displacement and other forms of coevolution is not impossible, as demonstrated by Levinton's (1982) laboratory tests of whether body-size differences between species of *Hydrobia* snails indicate different food particle sizes. Earlier work had assumed that body-size differences in these snails were equivalent to character displacement and that the differences allowed the species to avoid interspecific competition. Levinton's (1982) results are negative; the species show no indication of specializing on different particle sizes.

A pair of fascinating experimental studies with different lines of *Drosophila melanogaster* by Sulzbach and Emlen (1979) and Sulzbach (1980) was directed at the question of competitive coevolution. Do different lines of flies that are forced to compete intensely, at extraordinarily high densities, evolve any sort of interspecific tolerance or enhanced competitive ability? The results were negative. None of the populations significantly changed in competitive ability relative to control populations in the latter study (Sulzbach 1980), and the few significant changes in the earlier study (Sulzbach and Emlen 1979) did not reflect a change that could be attributed to the particular competing population.

Though not experimental, other recent empirical work relating to character displacement and competitive coevolution has come to similar negative conclusions: den Boer (1980) with carabid beetles; Huey (1979) with geckos; Simberloff (1984) with birds in general; Simberloff and Boecklen (1981) with many organisms. Although I am highly skeptical, other authors see character displacement in a positive light. For interpretations of character differences that tend to favor character displacement, see Brown and Bowers (1984) or Grant and Schluter (1984). V. C. Remsen (MS) has found that kingfishers meet some criteria for character-displaced species much better than many other organisms. These birds have food size differences among species that are correlated with morphological differences, and some species fish at identical sites and apparently draw food from the same source.

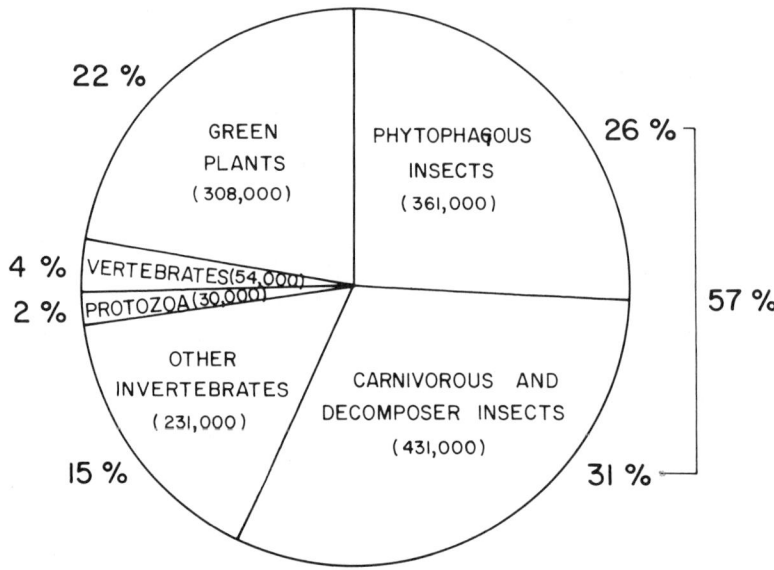

Fig. 1.—The approximate numbers (in parentheses) and proportions of different sorts of macroscopic species now on earth. Data from Southwood (1978).

PART 2

DIFFERENT ORGANISMS, DIFFERENT ECOLOGIES

Birds, lizards, and other vertebrates, which inspire much of orthodox competition theories in community ecology, are only a small fraction of biotic diversity, approximately 4% of extant macroscopic species (fig. 1). In contrast to the approximately 8,500 birds and 4,500 species of mammals now in existence, there are about one third of a million species of herbivorous insects.

HERBIVOROUS INSECTS

Herbivorous insects, with over one fourth of the macroscopic species on earth, lend little general support to theories of community ecology that assume strong inexorable competition between species, especially theories that are based on resource depletion (Lawton and Strong 1981; Strong et al. 1983). This point is not new (Hairston et al. 1960; Rathcke 1976) and was anticipated in early empirical studies with herbivorous insects that indicated a distinct lack of the simple, strong density dependence assumed by orthodox competition theory (Thompson 1928; Schwerdtfeger 1958; Milne 1957; Andrewartha and Birch 1954).

Density-Vague Population Ecology

Population change that is, at best, only weakly related to density is the norm for many herbivorous insects because of great effects of the weather, host plant phenology and physiology, and natural enemies that are not involved in negative feedback with the herbivores' population density (Strong 1983). I use the term density-vague to describe the normally low, even if sometimes significant, correlations between population parameters and density, and to avoid the implication that there is no relationship whatsoever between these variables. So much variance, discontinuity, and nonlinearity often cloud density relationships for populations of herbivorous insects and their natural enemies that any regulation or control is secondary to other factors affecting change, especially at usual densities. In these populations, general theories designed around the single linchpin of density are inappropriate. Various factors of autecology, food webs, and the weather explain most population change of herbivorous insects.

Density-vague population dynamics are not restricted to phytophagous insects. D. S. Wethy, R. S. Bowman, and J. R. Lewis (personal communication) have discovered that population change is not consistently related to density in barnacles, and suggest that other sessile organisms that are space-limited might behave in a similarly non-Malthusian fashion. Population parameters in fish populations may change in density-vague fashions, probably because critical actions of abiotic factors that are unrelated to stock size or density frequently affect recruitment greatly (Everhart and Young 1981; Laurence 1981). Survivorship of rainforest trees is density-vague according to the transplantation experiments of J. Connell (personal communication).

Interspecific competition has not been found to be an important general community mechanism for herbivorous insects. Most species do not suffer from severe depletion of resources, and neither exploitation nor interference is commonly shown to be a large factor in the coexistence of species. Overlap in use of resources does not usually imply interspecific interaction, and niche models do not account well for much of what we find for herbivorous insects, as is well summarized by a recent study that set out to test niche models for herbivorous insects: "Insect species are certainly not as equitably distributed over potential resources as a theory of species packing based on competition would lead one to expect" (Futuyma and Gould 1979, p. 47). This is not to say that no herbivorous insect communities have been interpreted in terms of orthodox competition theory (e.g., Denno 1980), but just that evidence for this particular community model is not commonly found among phytophagous insects. Even in situations of consistent resource depletion, as for caddis flies in high mountain streams, orthodox competition theories of the niche do not give much insight because instars of the same species are of different sizes, and because overlap in food particle sizes is not an index of interspecific competitive effect (Alstad 1982).

Interspecific competition is probably not commonly important for herbivorous insects because autecology, vertical food-web factors (from the host plant or from natural enemies), and the weather normally serve to maintain populations below densities that would deplete resources. Occasional outbreaks in some species, and rather common ones in other species, however, can lead to bouts of interspecific competition. External factors that rapidly reduce the availability of plant material might have an effect similar to greatly increased numbers of an insect species (Halkka 1978). Given the large literature on herbivorous insects, the paucity of studies showing strong interspecific competition is striking (Lawton and Strong 1981; Lawton and Hassell 1982).

Some cases of interspecific competition among herbivorous insects show how inconsequential this factor can be. One good example is given by Kareiva's (1982) experiments with collard-feeding insects, which he threw together into a cockpit of a host garden that maximized chances of intense species interactions. He discovered, however, that even with favorable homogeneous food resources and high insect densities, most opportunities for interspecific competition did not materialize. ". . . Most of the null comparisons between control and removal populations involved species pairs that lacked the opportunity to interact competitively . . . . *P. cruciferae* passed its peak abundance before the other two herbivores arrived in significant numbers . . . . Cabbage worms did not even appear until the 5th wk of experiment I; this provided very little opportunity for the caterpillars to inflict damage on the collards (or consequently the flea beetles) . . . ."(p. 701).

This study shows how a combination of the insects' autecology and the plant's phenology is crucial to the potential for interspecific competition. Lack of their coordination in the rapid phenology prevents much serious interaction between these herbivores.

Abiotic influences are often of extraordinary importance in communities of herbivorous insects. The weather and its results, such as fire (Whelan and Main 1979) and water stress to host plants (White 1976), determine a large fraction of structure and its variation for insects on plants. Good examples of weather's influence are a staple of empirical studies with phytophagous insects (e.g., Blau 1980; Dempster and Hall 1980; Ehrlich et al. 1972; Singer and Ehrlich 1979).

The biotic influences most important in communities of phytophagous insects are usually vertical, from above and below in the food web, rather than horizontal, from competitors on the same trophic level. The influences from below in the food web, from plant(s), are profound. Plant phenology (Mooney et al. 1981), plant nutrition for herbivores (Myers 1980), the noxious chemistry and physical attributes of plants (Rosenthal and Janzen 1979), plant density (their concentration as a resource; Risch 1981; Root 1973), the matrix of other plant species (Root 1973), and the influence of plants upon natural enemies of herbivorous insects (Atsatt and O'Dowd 1976) all have influences that usually far overshadow any influence of interspecific competition in these communities.

DECOMPOSER INSECTS

A second quarter of extant species on earth comprises insects that feed, at least partially, on dead or dying tissue and insects that are predators or parasites (fig. 1). Fewer community studies have been done with decomposers than with herbivores, and several studies have concerned decomposers together with other trophic classes of insects. One set of studies that stands out for attention to critical assessment of the role of competition among intimately associated insects is that of Seifert and Seifert, done in bracts of *Heliconia* plants in tropical America (Seifert 1984). The first study (Seifert and Seifert 1976) dealt with two species of beetles, two species of flies, and a cockroach that inhabit the water-holding bracts of *Heliconia wagneriana* and *H. imbricata* in Costa Rica. Of all of the possible pairwise interactions on the two plant species, 13 were not statistically significant, five were mutualistic (the opposite of competitive), and only four were competitive. On *H. wagneriana* the mutualistic interaction had greater intensity than either of the two competitive interactions. On *H. imbricata,* three of the four mutualistic interactions were more intense than either of the competitive interactions. Their second study, done at high elevation in Venezuela in the bracts of *H. bihai* (Seifert and Seifert 1979), found a similar lack of interspecific competitive organization for the community. The overwhelming effect upon these insects was from plant phenology, a vertical food web influence, rather than the horizontal influence between potential competitors assumed by competition theory.

Another community study with decomposer insects (Wise 1981) addressed the problem of net transgenerational effects of any interspecific competition, with ground beetles. From 1975 to 1978 Wise removed the most abundant ground beetle species, which comprised more than 50% of the biomass. Adult population densities of the other four ground beetle species did not increase in response to reduction of the most abundant species. Nor was there any consistent effect on the size of eclosing adults of the other species, which would have indicated some sort of "competitive release."

Other work with mycetophagous insects, which live beneath bark and might be loosely grouped into the category of decomposer insects, gives no indication that these communities are structured, or even much affected by competition (Taylor 1980).

Insects feeding upon decaying animal tissue (carrion) often subsist on a sequence of short-lived patches of food that change both as a result of physical influences such as drying and as a result of other kinds of decomposers such as microbes or vertebrates. Beaver (1979) has compared insects using carrion and dead petioles of Malaysian plants. In neither case did competition result in even the hint of exclusion of a species or suggest a model of this community analogous to niche theory. Overlap was very high, but interspecific interaction probably was not great, because each population is continually broken into small segments in discrete bits of the resources (individual animal carcasses or dying plant parts) and any competition is extremely localized. Similarly, dung beetles occupy sequences of transient food-habitat patches. Holter (1982) found that 13 species of dung beetles in the genus *Aphodius* coexist without depleting more than a few percent of available resources and without much evidence of any competition, past or present.

The theoretical model that most accurately describes these organisms is not one of homogeneous niches at some equilibrium but one with highly dissected, rapidly changing resource patches (Atkinson and Shorrocks 1981; Levin 1976; Slatkin 1974).

PREDATORY AND PARASITIC INSECTS

Some work with these organisms does indicate interspecific competition, albeit not competition that is generally suggestive of niche models. In some cases, with parasitoids, overlap does seem to indicate the potential for competition, and competitive exclusion has been observed in cases of biological control (DeBach 1974).

The parasitoids of the gall midge *Rhopalomyia californica* on coyote brush, *Baccharis pilularis,* are among the most studied group of this sort, and information from this guild indicates how wrong a naive niche model would be for these organisms. Force (1974) showed that these parasitoids would compete intensely in the laboratory, resulting in patterns of relative abundance very different from those found in nature. From his work it was clear that competition in a uniform, unchanging matrix of host insects was not what was occurring in nature; polyphagous hyperparasitoids probably lower the emergence rate of the most successful competitors. But hyperparasitism is not all of the story. Colonizing and host-finding ability differs between species and is responsible for much of the pattern of relative abundance in nature (Hopper 1983).

Other recent studies with *Rhopalomyia* on urban populations of the coyote brush, by Ehler (1982), indicate how the community effects of interspecific competition cannot be deduced beforehand for parasitoids. Each parasitoid species has a distinct autecology and phenology, and in the urban habitat relative abundance is not dominated by interspecific competition. Any competition between particular pairs of species varies with the seasons, and a large fraction of each species escapes the effects of competitors. Ehler (1978) has shown how additional species can lead to rates of pest destruction higher than those caused by a single species, a result at variance with inferences based upon deductions from much theory that assumes dynamics of parasitoids are dominated by competition.

Ehler's (1978) work builds upon the classical study of Huffaker and Kennett (1966), who found a distinctly complementary influence between two parasitoid species in biocontrol of the olive scale. One parasitoid species is more active in the spring; the other is more active in the fall. The autecologies of the two are so different that adverse weather that decreased performance of one usually did not badly affect the other; in combination, these parasitoid species, though partial competitors, buffer biological control of the olive scale. Such discoveries are based on hard-nosed empiricism. They have not relied on niche theory, and they tend to indicate how simplistic purely competitionist deductions can be.

Predaceous insects, like parasitoids, may compete between species (Pearson and Mury 1979); in many cases, however, competition does not overwhelm the ecology of these organisms either. A classic example derives from Johnson (1910, p. 87), who made the earliest known use of the term "niche" (Gaffney 1975; Hutchinson 1978). On the basis of his study of ladybird beetles, Johnson rejected the "Malthusian" deduction that competition between similar species invariably has severe results, and he rejected the importance of resource depletion.

Spiders are arthropod predators that have been interpreted variously in terms of interspecific competition. Wise (1984) reviewed the literature on spider communities, and points out how several authors have inferred orthodox niche relations and competitive organization from indirect observational evidence, on the basis of patterns of spatial distribution, relative abundance, diurnal activity, phenology, and habitat use. On the other hand, similar patterns examined more closely have produced skeptical interpretations. For example, Gertsch and

Riechert (1976), Post and Riechert (1977), Turner and Polis (1979), and Maelfait et al. (1980) inferred that interspecific competition is, at most, of secondary importance for most spiders. One striking result of these studies is that "niche" overlap does not indicate interaction; both high and low overlap can be evidence for or against competition, depending upon the assumptions.

Certainly intraspecific competition and cannibalism are important in spider communities, but these often do not translate into commensurate interspecific effects (Schaefer 1975, 1978; Wise 1979, 1981; Riechert and Cady, in press). As Wise (1984) emphasizes, "Clear experimental evidence of major competitive interactions between closely related spider species is lacking. Removal and density-manipulation experiments have not established interspecific competition to be a significant interaction in spider communities."

OTHER ORGANISMS

I work primarily with insects on plants, but cannot help finding in literature on other organisms indications of the principles that have emerged from entomological studies with herbivores. For at least one community of bactivorous ciliate protozoa, density-vague population dynamics and individualistic species coexistence appear to occur (Taylor 1979). For many invertebrates, vertebrates, and even green plants, autecology, the weather, and food web relations between trophic levels have profound influence in communities. Competition between species sometimes is detectable, but its influence can be minor in many cases relative to those more primary ecological forces. As well, when competition does occur, it often takes the form of interference rather than the exploitation assumed by many niche analyses.

Communities of invertebrate parasites other than insects have been reviewed recently with the distinct conclusion, "There is little evidence to suggest that competition has been an organizing force in the studies indicating nonoverlapping, noninteractive coexistence of parasites" (Price 1980, p. 139). One important contribution to the young but growing literature on factors other than interspecific competition in communities is that of Klaus Rohde, with monogenean parasites of fish. Rohde (1979) argues that "only a small portion of the niches available to ectoparasites of fish is filled" (p. 667), and "parasite species with coinciding or overlapping microhabitats often show no interactions" (p. 667). The evidence provided by these parasites suggests that autecological factors such as aggregation for the purposes of finding mates and physiological influences from the host fish are of prime importance in these communities. Many interspecific influences among species are likely to be positive rather than competitive for these parasites.

Andrews and Petney (1981) have investigated the coexistence of reptile ticks with results similar to those of Price (1980) and Rohde (1979). Their research turned up little evidence of interspecific competition for attachment sites on hosts or evidence that such competition in the past or present maintains the parapatric boundaries of geographical distribution for these organisms.

Certainly not all parasites behave in this noninteractive fashion, as Price (1984) and Holmes (1979) point out; this diversity of community mechanisms, even within taxa, is a major point of my paper. In shallow California waters, several species of digenetic trematodes so thoroughly infect populations of some marine snails that interspecific competition is very likely (A. Kuris and W. Sousa, personal communication). From empirical studies, Kuris and Sousa have calculated competitive hierarchies between species of these trematodes. The high incidences of parasitism by the California trematodes are in great contrast to the lower incidences by monogenetic ones that Rohde (1979) treats, and herein lies one key to understanding the differences between communities in degree and intensity of competitive interactions. The Kuris-Sousa findings contrast greatly with the incidence of snail parasitism by monogenetic trematodes on the Gulf coast of Florida, where one extensive study found only 15.2% infection from all species combined (Holliman 1961). As pointed out by Price (1980), low overall rates of host occupation by parasites result in very low rates of interspecific co-occurrence quite independently of any interaction among species. Learning about the causes of differences between California and Gulf coast existence for monogenean trematodes would be a great contribution to community ecology.

Green plants are approximately one fourth of present species (fig. 1). These sessile organisms commonly cover their substrates so thoroughly that some interspecific competition appears to be a reasonable expectation at the outset, especially in contrast to many of the other organisms that I have discussed above; but to assume that interspecific competition is the overwhelming factor in the coexistence of plant species in general is to ignore the same primary forces of autecology that must be considered for insects and other animals. For example, autecological adaptations of plants to environmental gradients may explain zonation of some plants without having to invoke competition at all. Mangroves illustrate this argument. Their seeds settle and grow in relation to the physical depth of water at high tide, with little evidence of competitive sorting between species (Rabinowitz 1978). Disturbances, disruptions, and stress are such prominent features of plant environments that one theorist (Grime 1977) has suggested that only in relatively productive, undisturbed vegetation should competitive forces prevail over autecological ones. Moreover, even in plant communities where interspecific competition has been shown to be important (such as mown fields of the temperate zone), high levels of uncertainty, nonlinear interactions, weak, diffuse competition, and intricate spatial interdispersions (Fowler 1981) suggest that the simple, deterministic niche models of ideal intense competition at equilibrium between pairs of species do not describe community behavior. A. Shmida and S. Ellner (in press) and A. Shmida and M. V. Wilson (personal communication) give formal models of this sort of spatially and temporally disjunct community existence.

Niche models of competition do not account well for at least one more facet of existence and coexistence of plants, sparseness. Some sparse plant species are always locally rare, even though they may have wide geographical distributions. Rabinowitz (1981) suggests that sparse plants have no "favored" habitat; they do not fall into a position along a simple gradient of niche analysis, nor should they have a fixed position in a competitive hierarchy. Their summed competitive effect on other species in the community is probably very small, even though individuals may compete quite effectively; certainly poor competitive abilities of sparse plant species cannot be the cause of their sparseness, as would be assumed by a naive competition theory.

The apparently nonequilibrium nature of many plant communities, especially at low latitudes where most species occur (Hartshorn 1978, 1980; Connell 1978; Grime 1979), leads one to theoretical interpretations very different from those of orthodox competition theory (Connell 1979; Caswell 1982). Even at high latitudes, forests apparently have been greatly disturbed, repeatedly and over long periods of time (Henry and Swan 1974; Auclair and Goff 1971). Fire (Wright 1974) and man are the two most obvious sources of continual disturbance to high-latitude forests, but insects may be a source of disturbance that is both unappreciated and important (McLeod 1980). Balsam fir, for example, is very susceptible to fire and regenerates poorly after being burned. The spruce budworm facilitates regeneration of this tree by creating light gaps in the forest by defoliation without disturbing the soil. Without budworm, balsam fir might be restricted to regenerating in treefall gaps or gaps left by the death of isolated old trees; the insect's defoliations may afford this tree its great abundance in Canadian forests (McLeod 1980; Blais 1965; Baltzer 1969).

Vertebrate communities are the archetypal model for orthodox competition theory (MacArthur 1972; Lack 1971; Cody 1974; Diamond 1978), but even in this royalist province, Whiggery is increasingly evident. In the empirical literature about communities of vertebrates substantial questions about the ubiquity and relative importance of competition have recently been raised. Now, to be sure, physical aggressiveness of some birds, especially nectivores (Wolf 1978), and rodents results in interspecific displacement (interference) (Brown 1971; Grant 1972), but whether interspecific competition generally overwhelms autecological, food-web, and weather influences for vertebrates is appearing increasingly problematical.

Tinkle (1982, p. 57) provides a good example concerning a diverse community of sceloporine lizards, in Arizona. His work was experimental, covered several generations, and was based upon a sound understanding of the autecology of these lizards. Tinkle removed two of the three most abundant species, with the result of "no detectable effect of the removal . . . on habitat selection, perch height, survivorship, population density, or individual body size" of the remaining species under investigation. The lack of any significant, detected, competitive effect between species has to be qualified. Other detailed experimental studies with these desert lizards (Dunham 1980, 1981) have detected significant, if sometimes small and usually transitory, effects of interspecific competition. Some variables (e.g., adult survival) responded to the experiments in a fashion opposite to that predicted by competition theory, and environmental variability was clearly a major influence for these organisms. Competitive effects vary greatly among years. Other studies reinforce the conclusion of high natural variability in diet and diet overlap among species of these lizards (Vitt et al. 1981).

Schoener and Schoener (1982) provide another experimental study of lizard (*Anolis*) communities that reaffirms the importance of factors other than competition in vertebrate communities. They found that predation by birds, perhaps together with diffuse competition from birds, greatly affected survival of *Anolis* lizards. Adolph and Roughgarden (1983) have noted the potential influence of predation by birds on *Anolis*. The weather has substantial influence on ecological success of *Anolis* lizards (Stamps and Tanaka 1981), and these findings lead one to wonder if lizard communities are generally all that different from communities of herbivorous insects, where interspecific competition plays a minor role in coexistence of species. This concern is reflected in a recent study of *Anolis* that found "low energy demands, a relative abundance of suitable prey sizes in the environment, and low search costs may allow such animals to feed more like herbivores than insectivorous birds or mammals" (Stamps et al. 1981, p. 1,079). I appreciate the evidence for intense present-day competition between some co-occurring *Anolis* species (Pacala and Roughgarden 1982). Because we now know the importance of several factors, biotic and abiotic, in these communities, theory, observation, and experiments will be quite focused in the future and the potential for advancement in understanding is great.

ACKNOWLEDGMENTS

Special thanks are due to Gary Grossman and I. J. Schlosser for their insightful comments on the manuscript. I also thank A. Dunham, J. Roughgarden, N. Gotelli, L. Abele, F. James, J. Connell, and D. Simberloff for their input, and Si Levin for a brief but informative chat about the symbiosis of ecological theory and empiricism.

LITERATURE CITED

Abrams, P. 1976. Limiting similarity and the form of the competition coefficient. Theor. Popul. Biol. 8:356–375.

Adolph, S., and J. Roughgarden. 1983. Foraging by passerine birds and *Anolis* lizards on St. Eustatius (Neth. Antilles): implications for interclass competition and predation. Oecologia 56:313–317.

Aho, J., E. Ranta, and J. Vourinen. 1981. Species composition of freshwater snail communities in lakes of south and west Finland. Ann. Zool. Fenn. 18:233–242.

Alstad, D. N. 1982. Current speed and filtration rate link caddisfly phylogeny and distributional patterns on a stream gradient. Science 216:533–534.

Andrewartha, H. G., and L. C. Birch. 1954. The distribution and abundance of animals. University of Chicago Press, Chicago.

Andrews, R. H., and T. N. Petney. 1981. Competition for sites of attachment to hosts in three parapatric species of reptile tick. Oecologia 51:227–232.

Armstrong, R. A., and R. McGehee. 1980. Competitive exclusion. Am. Nat. 115:151–170.

Arthur, W. 1981. The evolutionary consequences of interspecific competition. Adv. Ecol. Res. 13:127–187.

Atkinson, W. D., and B. Shorrocks. 1981. Competition on a divided and ephemeral resource: a simulation model. J. Anim. Ecol. 50:461–471.

Atsatt, P. R., and D. J. O'Dowd. 1976. Plant defense guilds. Science 193:24–29.

Auclair, A. N., and F. G. Goff. 1971. Diversity relations of upland forests in the western Great Lakes area. Am. Nat. 105:499–528.

Baltzer, H. O. 1969. Forest character and vulnerability of balsam fir to spruce budworm in Minnesota. For. Sci. 15:17–25.

Beaver, R. A. 1979. Non-equilibrium "island" communities. A guild of tropical bark beetles. J. Anim. Ecol. 48:987–1002.

Berkeley, G. 1713. Three dialogues between Hylas and Philonous. *In* A. A. Luce and T. E. Jessop, eds. Works. Vol. 1. Philosophical commentaries. Essay toward a new theory of vision. Theory of vision vindicated and explained. T. Nelson, London, 1948.

Birch, L. C. 1979. The effect of species of animals which share common resources on one another's distribution and abundance. Fortschr. Zool. 25:197–221.

Blais, J. R. 1965. Spruce budworm outbreaks in the past three centuries in the Laurentide Park, Quebec. For. Sci. 11:130–138.

Blau, W. S. 1980. The effect of environmental disturbance on a tropical butterfly population. Ecology 61:1005–1012.

Boag, P. T., and P. R. Grant. 1981. Intense natural selection on a population of Darwin's finches (Geospizinae) in the Galápagos. Science 214:82–85.

Boulding, K. E. 1980. Science: our common heritage. Science 207:831–836.

British Ecological Society. 1944. Symposium on the ecology of closely allied species. J. Anim. Ecol. 13:176–178.

Brown, J. H. 1971. Mechanisms of competitive exclusion between two species of chipmunks (*Eutamias*). Ecology 52:306–311.

Brown, J. H., and M. A. Bowers. 1984. Patterns and processes in three guilds of terrestrial vertebrates. *In* D. R. Strong, Jr., D. Simberloff, L. G. Abele, and A. B. Thistle, eds. Ecological communities: conceptual issues and the evidence. Princeton University Press, Princeton, N.J. (in press).

Buss, L. W. 1981. Group living, competition, and the evolution of cooperation in a sessile invertebrate. Science 213:1012–1014.

Buss, L. W., and J. B. C. Jackson. 1979. Competitive networks: nontransitive competitive relationships in cryptic coral reef environments. Am. Nat. 113:223–234.

Caswell, H. 1982. Life history and the equilibrium status of populations. Am. Nat. 120:317–339.

Chase, S. 1938. The tyranny of words. Harcourt, Brace, New York.

Chesson, P. 1978. Predator-prey theory and variability. Annu. Rev. Ecol. Syst. 9:323–347.

Cody, M. L. 1974. Competition and the structure of bird communities. Princeton University Press, Princeton, N. J.

Connell, J. H. 1961. The influence of interspecific competition and other factors on the distribution of the barnacle *Chthalamus stellatus*. Ecology 42:710–723.

———. 1975. Some mechanisms producing structure in natural communities: a model and evidence from field experiments. Pages 460–490 *in* M. L. Cody and J. M. Diamond, eds. Ecology and evolution of communities. Harvard University Press, Cambridge, Mass.

———. 1978. Diversity in tropical rain forests and coral reefs. Science 199:1302–1310.

———. 1979. Tropical rain forests and coral reefs as open nonequilibrium systems. Symp. Br. Ecol. Soc. 20:141–163.

———. 1983. On the prevalence and relative importance of interspecific competition: evidence from field experiments. Am. Nat. 122:661–696.

Dayton, P. K. 1971. Competition, disturbance, and community organization: the provision and subsequent utilization of space in a rocky intertidal community. Ecol. Monogr. 41:351–389.

DeBach, P. 1974. Biological control by natural enemies. Cambridge University Press, Cambridge.

Dempster, J. P., and M. L. Hall. 1980. An attempt at re-establishing the swallowtail butterfly at Wicken Fen. Ecol. Entomol. 5:327–334.

den Boer, P. J. 1980. Exclusion or coexistence and the taxonomic or ecological relationship between species. Neth. J. Zool. 30:278–306.

Denno, R. F. 1980. Ecotope differentiation in a guild of sap-feeding insects on the salt marsh grass, *Spartina patens*. Ecology 61:702–714.

Diamond, J. M. 1978. Niche shifts and the rediscovery of interspecific competition. Am. Sci. 66:322–331.

Downhower, J. F. 1976. Darwin's finches and the evolution of sexual dimorphism in body size. Nature 263:558–563.

Dunham, A. E. 1980. An experimental study of interspecific competition between the iguanid lizards *Sceloporus merriami* and *Urosaurus ornatus*. Ecol. Monogr. 50:309–330.

———. 1981. Realized niche overlap, resource abundance and the intensity of interspecific competition in a guild of iguanid lizards. *In* Proceedings of a symposium on the ecology of lizards. American Society of Zoologists meetings, Seattle, December 1980.

Dunham, A. E., D. W. Tinkle, and J. W. Gibbons. 1978. Body size in island lizards: a cautionary tale. Ecology 59:1230–1238.

Ehler, L. E. 1978. Competition between two natural enemies of Mediterranean black scale on olive. Environ. Entomol. 7:521–523.

———. 1982. Ecology of *Rhopalomyia californica* Felt (Diptera: Cecidomyiidae) and its parasites in an urban environment. Hilgardia 50:1–32.

Ehrlich, P. P., D. E. Breedlove, P. F. Brussard, and M. A. Sharp. 1972. Weather and the "regulation" of subalpine populations. Ecology 53:243–247.

Everhart, W. H., and W. O. Young. 1981. Principles of fisheries science. Comstock, Ithaca, N.Y.

Fleischer, R. C., and R. F. Johnston. 1982. Natural selection on body size and proportions in house sparrows. Nature 298:747–749.

Force, D. C. 1974. Ecology of insect host-parasitoid communities. Science 184:624–632.

Fowler, N. 1981. Competition and coexistence in a North Carolina grassland. II. The effects of the experimental removal of species. J. Ecol. 69:843–854.

Futuyma, D. J., and F. Gould. 1979. Associations of plants and insects in a deciduous forest. Ecol. Monogr. 49:33–50.

Gaffney, P. M. 1975. Roots of the niche concept. Am. Nat. 109:490.

Gertsch, W. J., and S. E. Riechert. 1976. The spatial and temporal partitioning of a desert spider community with descriptions of new species. Am. Mus. Novit. 2604:1–25.

Getz, W. M., and A. P. Gutierrez. 1982. A perspective on systems analysis in crop production and insect pest management. Annu. Rev. Entomol. 27:447–466.

Ghiselin, M. T. 1969. The triumph of the Darwinian method. University of California Press, Berkeley.

Gleason, H. A. 1926. The individualistic concept of the plant association. Bull. Torrey Bot. Club 53:7–26.

Grant, P. R. 1972. Interspecific competition among rodents. Annu. Rev. Syst. Ecol. 3:79–106.

Grant, P., and D. Schluter. 1984. Interspecific competition inferred from patterns of guild structure. *In* D. R. Strong, Jr., D. Simberloff, L. G. Abele, and A. B. Thistle, eds. Ecological communities: conceptual issues and the evidence. Princeton University Press, Princeton, N.J. (in press).

Grime, J. P. 1977. Evidence for the existence of three primary strategies in plants and its relevance to ecological and evolutionary theory. Am. Nat. 111:1169–1194.

———. 1979. Plant strategies and vegetation processes. Wiley, London.

Grossman, G. D., P. B. Moyle, and J. O. Whitaker. 1982. Stochasticity in structural and functional characteristics of an Indiana stream fish assemblage: a test of community theory. Am. Nat. 120:423–454.

Hairston, N. G. 1981. An experimental test of a guild: salamander competition. Ecology 62:65–72.

Hairston, N. G., F. E. Smith, and L. B. Slobodkin. 1960. Community structure, population control, and competition. Am. Nat. 94:421–425.

Halkka, O. 1978. Influence of spatial and host-plant isolation on polymorphism in *Philaenus spumarius*. Pages 41–55 *in* L. A. Mound and N. Waloff, eds. Diversity of insect faunas. Symp. R. Entomol. Soc. Lond. 9, Blackwell, Oxford.

Hartshorn, G. S. 1978. Treefalls and tropical forest dynamics. Pages 617–638 *in* P. B. Tomlinson and M. H. Zimmermann, eds. Tropical trees as living systems. Cambridge University Press, Cambridge.

———. 1980. Neotropical forest dynamics. Biotropica 12 (suppl.):23–30.

Henry, J. D., and J. M. A. Swan. 1974. Reconstructing forest history from live and dead plant material and an approach to the study of forest succession in southwest New Hampshire. Ecology 55:772–783.

Hespenheide, H. A. 1971. Food preference and the extent of overlap in some insectivorous birds with special reference to the Tyrannidae. Ibis 113:59–72.

———. 1973. Ecological inferences from morphological data. Annu. Rev. Ecol. Syst. 4:213–229.

———. 1975. Prey characteristics and predator niche width. Pages 158–180 *in* M. L. Cody and J. M. Diamond, eds. Ecology and evolution of communities. Harvard University Press, Cambridge, Mass.

Holliman, R. B. 1961. Larval trematodes from the Apalachee Bay area, Florida, with a checklist of known marine cercariae arranged in a key to their superfamilies. Tulane Stud. Zool. 9:2–74.
Holmes, J. C. 1979. Parasite populations and host community structure. Pages 27–46 in B. B. Nichol, ed. Host-parasite interfaces. Academic Press, New York.
Holter, P. 1982. Resource utilization and local coexistence in a guild of scarabaeid dung beetles (*Aphodius* spp.). Oikos 39:213–227.
Hopper, K. 1983. Determinants of abundance in a guild of parasitic wasps: host finding and colonizing ability. Ecology (in press).
Huey, R. B. 1979. Parapatry and niche complementarity of Peruvian desert geckos (*Phyllodactylus*): the ambiguous role of competition. Oecologia 38:249–259.
Huffaker, C. B., and C. E. Kennett. 1966. The biological control of *Parlatoria oleae* (Colvée) through the compensatory action of two introduced parasites. Hilgardia 37:283–334.
Hurlbert, S. H. 1978. The measurement of niche overlap and some relatives. Ecology 59:67–77.
Hutchinson, G. E. 1978. An introduction to population ecology. Yale University Press, New Haven, Conn.
Jaeger, R. C. 1974. Interference or exploitation? A second look at competition between salamanders. J. Herpetol. 8:191–194.
James, F. C. 1970. Geographic size variation in birds and its relationship to climate. Ecology. 51:365–390.
———. 1982. The ecological morphology of birds: a review. Ann. Zool. Fenn. 19:265–275.
Johnson, R. H. 1910. Determinate evolution in the color pattern of the lady-beetles. Carnegie Inst. Wash. Publ. 122.
Kareiva, P. 1982. Exclusion experiments and the competitive release of insects feeding on collards. Ecology 63:696–704.
Keen, W. H. 1982. Habitat selection and interspecific competition in two species of Plethodontid salamanders. Ecology 63:94–102.
Krebs, C. J. 1978. Ecology: the experimental analysis of distribution and abundance. 2d ed. Harper & Row, New York.
Lack, D. 1947. Darwin's finches. Cambridge University Press, Cambridge.
———. 1971. Ecological isolation in birds. Harvard University Press, Cambridge, Mass.
Laurence, G. C. 1981. Overview—Modelling—An esoteric or potentially utilitarian approach to understanding larval fish dynamics? Pages 3–7 in R. Lasker and K. Sherman, eds. The early life history of fish: recent studies. Second ICES Symposium. Conseil International pour l'Exploration de la Mer. Palaegade 2–4, DK-1261 Copenhague K, Danemark, October 1981.
Lawton, J. H., and M. P. Hassell. 1982. Interspecific competition in insects. In C. B. Huffaker and R. L. Rabb, eds. Ecological entomology. Wiley, New York (in press).
Lawton, J. H., and D. R. Strong. 1981. Community patterns and competition in folivorous insects. Am. Nat. 118:317–338.
Levin, S. A. 1976. Population dynamic models in heterogeneous environments. Annu. Rev. Ecol. Syst. 7:287–310.
———. 1981. The role of theoretical ecology in the description and understanding of populations in heterogeneous environments. Am. Zool. 21:865–875.
Levins, R. 1968. Evolution in changing environments. Princeton University Press, Princeton, N. J.
Levinton, J. S. 1982. The body size-prey size hypothesis: the adequacy of body size as a vehicle for character displacement. Ecology 63:869–872.
Lewis, C. S. 1961. Sense. Pages 146–156, IX–X in Studies in words. Cambridge University Press, Cambridge.
Likens, G. F., F. H. Borman, R. S. Pierce, J. S. Eaton, and N. M. Johnson. 1977. Biochemistry of a forested ecosystem. Springer-Verlag, New York.
Lindeman, R. L. 1942. The trophic-dynamic aspect of ecology. Ecology 23:399–418.
MacArthur, R. H. 1971. Patterns of terrestrial bird communities. Pages 189–221 in D. S. Farner and J. R. King, eds. Avian biology. Vol. 1. Academic Press, New York.
———. 1972. Geographical ecology: patterns in the distribution of species. Harper & Row, New York.
McGowan, J. A., and P. W. Walker. 1979. Structure in the copepod community of the North Pacific central gyre. Ecol. Monogr. 49:195–226.

McIntosh, R. P. 1975. H. A. Gleason—"Individualistic Ecologist" 1882–1975: his contributions to ecological theory. Bull. Torrey Bot. Club 102:253–273.

———. 1980. The background of some current problems of theoretical ecology. Pages 1–62 *in* E. Saarinen, ed. Conceptual issues in ecology. Reidel, Boston.

McLeod, J. M. 1980. Forests, disturbances, and insects. Can. Entomol. 112:1185–1192.

Maelfait, J. P., L. Baert, J. Hublé, and A. De Kimpe. 1980. Life cycle timing, microhabitat preference and coexistence of spiders. Pages 69–73 *in* J. Gruber, ed. Proc. 8th Int. Arachnol. Congr. Vienna, 1980.

May, R. M. 1981. The role of theory in ecology. Am. Zool. 21:903–910.

May, R. M., J. R. Beddington, C. W. Clark, S. J. Holt, and R. M. Laws. 1979. Management of multispecies fisheries. Science 205:267–277.

Milne, A. 1957. The natural control of insect populations. Can. Entomol. 89:193–213.

Mooney, H. A., K. S. Williams, D. E. Lincoln, and P. R. Ehrlich. 1981. Temporal and spatial variability in the interaction between the checkerspot butterfly, *Euphydryas chalcedonia* and its principal food source the California shrub *Diplacus aurantiacus*. Oecologia 50:195–198.

Murdoch, W. W. 1979. Predation and the dynamics of prey populations. Fortschr. Zool. 25:295–310.

Myers, J. H. 1980. Is the insect or the plant the driving force in the cinnabar moth-tansy ragwort system? Oecologia 47:16–21.

Oster, G. 1975. Stochastic behavior of deterministic models. Pages 24–37 *in* S. A. Levin, ed. Ecosystem analysis and prediction. Society for Industrial and Applied Mathematics, Philadelphia.

Pacala, S., and J. Roughgarden. 1982. Resource partitioning and interspecific competition in two-species insular *Anolis* lizard communities. Science 217:444–446.

Paine, R. T. 1966. Food web complexity and species diversity. Am. Nat. 100:65–75.

Pavitrananda, S. 1944. Common sense about Yoga. Advaita Ashrama, Calcutta.

Peabody, R. R. 1931. The common sense of drinking. Little, Brown, Boston.

Pearson, D. L., and E. J. Mury. 1979. Character divergence and convergence among tiger beetles (Coleoptera: Cicindellidae). Ecology 60:557–566.

Peters, R. H. 1976. Tautology in evolution and ecology. Am. Nat. 110:1–12.

———. 1980. Useful concepts for predictive ecology. Synthèse 43:257–269.

Poole, R. W. 1977. Periodic, pseudoperiodic, and chaotic population fluctuations. Ecology 58:210–213.

Popper, K. 1962. Conjectures and refutations: the growth of scientific knowledge. Basic Books, New York.

Portwood, D. 1978. Common sense suicide: the final right. Dodd, Mead, New York.

Post, W. M., III, and S. E. Riechert. 1977. Initial investigation into the structure of spider communities. J. Anim. Ecol. 46:729–750.

Price, P. W. 1980. Evolutionary biology of parasites. Princeton University Press, Princeton, N.J.

———. 1984. Communities of specialists: vacant niches in ecological and evolutionary time. *In* D. R. Strong, Jr., D. Simberloff, L. G. Abele, and A. B. Thistle, eds. Ecological communities: conceptual issues and the evidence. Princeton University Press, Princeton, N.J. (in press).

Pulliam, H. R. 1983. Ecological community theory and the coexistence of sparrows. Ecology 64:45–52.

Quinn, J. F., and A. E. Dunham. 1983. On hypothesis testing in ecology and evolution. Am. Nat. 122:602–617.

Rabinowitz, D. 1978. Early growth of mangrove seedlings in Panama and an hypothesis concerning the relationship of dispersal and zonation. J. Biogeogr. 5:113–134.

———. 1981. Seven forms of rarity. Pages 205–217 *in* H. Synge, ed. The biological aspects of rare plant conservation. Wiley, New York.

Rathcke, B. J. 1976. Competition and coexistence within a guild of herbivorous insects. Ecology 57:76–88.

Riechert, S. E., and A. B. Cady. 1983. Patterns of resource use and tests for competitive release in a spider community. Ecology 64 (in press).

Rigler, F. H. 1975. The concept of energy flow and nutrient flow between trophic levels. Pages 15–26 *in* W. H. van Dobben and R. H. Lowe-McConnell, eds. Unifying concepts in ecology. Junk, Wageningen.

Risch, S. J. 1981. Insect herbivore abundance in tropical monocultures and polycultures: an experimental test of two hypotheses. Ecology 62:1325–1341.

Rohde, K. 1979. a critical evaluation of intrinsic and extrinsic factors responsible for niche restriction in parasites. Am. Nat. 114:648–671.

Root, R. B. 1973. Organization of a plant-arthropod association in simple and diverse habitats: the fauna of collards (*Brassica oleracea*). Ecol. Monogr. 43:95–124.

Rosenthal, G. A., and D. H. Janzen, eds. 1979. Herbivores: their interaction with secondary plant metabolites. Academic Press, New York.

Roughgarden, J. 1983. Competition and theory in community ecology. Am. Nat. 122:583–601.

Schaefer, M. 1975. Experimental studies on the importance of interspecies competition for the lycosid spiders in a salt marsh. Pages 86–90 in Proc. 6th Int. Arachnol. Congr. Amsterdam, 1974.

———. 1978. Some experiments on the regulation of population density in the spider *Floronia bucculenta* (Araneida: Linyphiidae). Symp. Zool. Soc. Lond. 42:203–210.

Schlosser, I. J. 1982. Fish community structure and function along two habitat gradients in a headwater stream. Ecology 52:395–414.

Schluter, D. 1982. Distributions of Galápagos ground finches along an altitudinal gradient: the importance of food supply. Ecology 63:1504–1517.

Schmitt, R. J., and J. A. Coyer. 1982. The foraging ecology of sympatric marine fish in the genus *Embiotica* (Embiotocidae): importance of foraging behavior in prey size selection. Oecologia 55:369–378.

Schoener, T., and A. Schoener. 1982. The ecological correlates of survival in some Bahamian *Anolis* lizards. Oikos 39:1–16.

Schwerdtfeger, F. 1958. Is the density of animal populations regulated by mechanisms or by chance? 10th Int. Congr. Entomol. 4:115–122.

Seifert, R. P. 1984. Does competition structure communities: Field studies on neotropical *Heliconia* insect communities. *In* D. R. Strong, Jr., D. Simberloff, L. G. Abele, and A. B. Thistle, eds. Ecological communities: conceptual issues and the evidence. Princeton University Press, Princeton, N. J. (in press).

Seifert, R. P., and F. H. Seifert. 1976. A community matrix analysis of *Heliconia* insect communities. Am. Nat. 110:461–483.

———. 1979. A *Heliconia* insect community in a Venezuelan cloud forest. Ecology 60:462–467.

Shakespeare, W. The complete works. Edited by P. Alexander. Random House, New York. 1952.

Shapiro, A. M. 1975. The temporal component of butterfly species diversity. Pages 181–195 in M. L. Cody and J. M. Diamond, eds. Ecology and evolution of communities. Belknap, Cambridge, Mass.

Shmida, A., and S. Ellner. In press. Coexistence of plant species with similar niches. Vegetatio.

Simberloff, D. 1980. A succession of paradigms in ecology: essentialism to materialism and probabilism. Synthèse 43:3–39.

———. 1981. Community effects of introduced species. Pages 53–81 in M. H. Nitecki, ed. Biotic crises in ecological and evolutionary time. Academic Press, New York.

———. 1984. Properties of coexisting bird species in two archipelagoes. *In* D. R. Strong, Jr., D. Simberloff, L. G. Abele, and A. B. Thistle, eds. Ecological communities: conceptual issues and the evidence. Princeton University Press, Princeton, N. J. (in press).

Simberloff, D., and W. Boecklen. 1981. Santa Rosalia reconsidered: size ratios and competition. Evolution 35:1206–1228.

Singer, M. C., and P. R. Ehrlich. 1979. Population dynamics of the checkerspot butterfly *Euphydryas editha*. Fortschr. Zool. 25:53–60.

Slatkin, M. 1974. Competition and regional coexistence. Ecology 55:128–134.

———. 1980. Ecological character displacement. Ecology 61:163–178.

Southwood, T. R. E. 1978. The components of diversity. Symp. R. Entomol. Soc. Lond. 9:19–40.

Stamps, J., and S. Tanaka. 1981. The influence of food and water on growth rates in a tropical lizard (*Anolis aeneus*). Ecology 62:33–40.

Stamps, J., S. Tanaka, and V. V. Krishnan. 1981. The relationship between selectivity and food abundance in a juvenile lizard. Ecology 62:1079–1092.

Strong, D. R. 1980. Null hypotheses in ecology. Synthèse 43:271–286.

———. 1983. Density-vague population dynamics. *In* P. W. Price, C. N. Slobodchikoff, and W. S.

Gaud, eds. The new ecology: novel approaches to interactive systems. Wiley-Interscience, New York (in press).
Strong, D. R., J. A. Lawton, and T. R. E. Southwood. 1983. Community patterns and mechanisms of phytophagous insects. Blackwell, Oxford (in press).
Strong, D. R., Jr., L. A. Szyska, and D. S. Simberloff. 1979. Tests of community-wide character displacement against null hypotheses. Evolution 33:897–913.
Sulzbach, D. S. 1980. Selection for competitive ability: negative results in *Drosophila*. Evolution 34:431–436.
Sulzbach, D. S., and J. M. Emlen. 1979. Evolution of competitive ability in mixtures of *Drosophila melanogaster*: populations with an initial asymmetry. Evolution 33:1138–1149.
Taylor, V. A. 1980. Coexistence of two species of *Ptinella* Motschulsky (Coleoptera: Ptiliidae) and the significance of their adaptation to different temperature ranges. Ecol. Entomol. 5:397–411.
Taylor, W. D. 1979. Sampling data on the bactivorous ciliates of a small pond compared to neutral models of community structure. Ecology 60:876–883.
Thompson, W. R. 1928. A contribution to the study of biological control and parasitic introduction in continental areas. Parasitology 20:90–112.
Tinkle, D. W. 1982. Results of experimental density manipulation in an Arizona lizard community. Ecology 63:57–65.
Turner, M., and G. A. Polis. 1979. Patterns of co-existence in a guild of raptorial spiders. J. Anim. Ecol. 48:509–520.
Vitt, L. J., R. C. van Loben Sels, and R. D. Ohmart. 1981. Ecological relationships among arboreal desert lizards. Ecology 62:398–410.
Wang, Y. H., and A. P. Gutierrez. 1980. An assessment of the use of stability analyses in population ecology. J. Anim. Ecol. 49:435–452.
Whelan, R. J., and A. R. Main. 1979. Insect grazing and post-fire plant succession in south-west Australian woodland. Aust. J. Ecol. 4:387–398.
White, T. C. R. 1976. Weather, food and plagues of locusts. Oecologia 22:119–134.
Wiens, J. A. 1983. Avian community ecology: an iconoclastic view. *In* G. A. Clark and A. H. Brush, eds. Perspectives in ornithology. Cambridge University Press, Cambridge (in press).
Wiens, J. A., and J. T. Rotenberry. 1980. Patterns of morphology and ecology in grassland and shrubsteppe bird populations. Ecol. Monogr. 50:287–308.
Willson, M. S. 1971. Seed selection in some North American finches. Condor 73:415–429.
Wilson, D. S. 1975. The adequacy of body size as a niche difference. Am. Nat. 109:769–784.
Wise, D. H. 1979. Effects of an experimental increase in prey abundance upon the reproductive rates of two orb-weaving spider species. (Araneae: Araneidae). Oecologia 41:289–300.
———. 1981. A removal experiment with darkling beetles: lack of evidence for interspecific competition. Ecology 62:1107–1120.
———. 1984. The role of competition in spider communities: insights from field experiments with a model organism. *In* D. R. Strong, Jr., D. Simberloff, L. G. Abele, and A. B. Thistle, eds. Ecological communities: conceptual issues and the evidence. Princeton University Press, Princeton (in press).
Wolf, L. L. 1978. Aggressive social organization in nectivorous birds. Am. Zool. 18:765–778.
Wright, H. E. 1974. Landscape development, forest fires, and wilderness management. Science 186:487–495.

ON THE PREVALENCE AND RELATIVE IMPORTANCE OF INTERSPECIFIC COMPETITION: EVIDENCE FROM FIELD EXPERIMENTS

JOSEPH H. CONNELL

Department of Biological Sciences, University of California, Santa Barbara, California 93106

Submitted December 28, 1982; Accepted April 21, 1983

How much does present-day interspecific competition affect the distribution, abundance, and resource use of species in natural communities? This is a question that continues to engender controversy among ecologists. Ideally, it might be answered by measuring the influence of interspecific competition on a population relative to that of other relevant processes affecting it, such as the weather, predation, parasitism, mutualism, intraspecific competition, disturbances, etc. If a large number of such case studies existed we might then make a general evaluation of the relative importance of each process. In the absence of such information, I decided to examine some published studies of competition in an effort to address three questions: (1) How frequently does interspecific competition occur at present in nature; (2) what mechanisms account for the variability in its occurrence; (3) when it does occur, how strong is it compared to intraspecific competition?

To address these questions, I have taken a sample of published studies that used field experiments designed to detect interspecific competition. In previous reviews of field experimental evidence (Connell 1974, 1975), I have assessed the relative importance of competition, predation, and physical factors in community structure. Here I compare only the strengths of intraspecific and interspecific competition. This would have been difficult to do before now since field studies distinguishing them have begun to be published only recently. Although laboratory and greenhouse experiments designed to do this have been done for many years (de Wit 1960; see review in Harper 1977), to my knowledge the first controlled field experiments designed specifically to distinguish the two types of competition appeared in the past decade (e.g., Harger 1970*a*, 1970*b*, 1972; Wilbur 1972).

A comparison of the relative strengths of intraspecific and interspecific competition is interesting in several contexts. The first concerns the coexistence and abundance of competitors. If species A is superior to species B in interspecific competition, one might conclude that A will eventually eliminate B, unless something interrupts the process. If, however, intraspecific competition in A is stronger than interspecific competition on B, A may be self-limited below the density necessary to eliminate B. This was suggested as a possible mechanism promoting coexistence of grazing gastropods by Underwood (1978) and Creese and Underwood (1982). Some theoretical models also use the relative strengths to predict the possibility of stable coexistence of two competitors at equilibrium. Where competitors coexist, we might be interested in the relative roles of these processes in determining the density of both species. Another aspect concerns the partitioning of resources or habitats, in which the two types of competition act in opposing directions. In theory, increasing intraspecific competition should expand a species' niche whereas increased interspecific competition should reduce it.

Here I will confine my attention to evidence from controlled field experiments. This is not meant to imply that such experiments are always superior to nonexperimental evidence. In many circumstances experiments are neither feasible nor appropriate to test particular hypotheses. To address the questions posed in this paper, however, field experiments probably provide some of the best evidence available.

METHODS

The idea of using field experiments to study present-day competition is not new (Jackson 1981), but unless they are designed and executed so as to test relevant hypotheses (which many are not), they are of limited use. In fact, since field experiments are often regarded as the ne plus ultra of ecological research, poor ones can be worse than useless since the conclusions based on them are often accepted with little question. When field experiments were few the tendency was to use this small amount of information rather uncritically. Now there are plenty of field experiments to choose from and I feel that it is time to reexamine the evidence.

Birch (1957, p. 6) defines competition between animals as follows: "Competition occurs when a number of animals (of the same or of different species) utilize common resources, the supply of which is short; or if the resources are not in short supply, competition occurs when the animals seeking that resource nevertheless harm one or another in the process." (This definition should also apply to plants and microorganisms.) In most field experiments, the degree of resource competition or interference is experimentally manipulated by changing the population densities of the competitors.

The Design of Field Experiments to Measure Competition

A common sort of field experiment involves changing the abundance of one possible competitor, species A, and comparing the response of the other, species B, to its behavior in an unmanipulated control; this design is called type 1 in the studies listed in Appendix A. The response measured is usually (1) a change in density; (2) a change in some rate that could affect density, e.g., fecundity, growth, physiological condition, mortality, etc.; or (3) a niche shift, e.g., a change in type of resource used, microhabitat occupied, etc. While this experiment will detect interspecific competition, for the first two responses it is open to the alternative interpretation that the same or a greater response might have been seen in species B with a similar change in its own density. In other words, intraspecific competition might be equal to or stronger than interspecific.

To distinguish the two types of competition, a different design is required. The density of each species needs to be varied while either keeping the density of the other species constant or removing it entirely. The experimental design depends upon whether the species can be transplanted and/or enclosed without ill effects. If so, densities can be both increased and decreased from the ambient condition within the same experiment. This has the advantage that it reveals the density above which interspecific and intraspecific competition begin to occur. For example, Underwood (1978) caged grazing snails both in single-species and mixed-species treatments at a range of densities above and below average, all within a single experiment.

If, on the other hand, the species cannot be transplanted or enclosed without ill effects, densities can only be reduced, not increased. A possible design is to reduce one of the species to a series of lower densities while either leaving the second species unchanged or removing it entirely. Simultaneously the second species is reduced with the first unchanged or removed. While this design will also measure the strengths of both intraspecific and interspecific competition it is less powerful than the first, since it does not include densities in the upper part of the natural range, at which competition is more likely to occur. However, it is worth doing to measure both sorts of competition; I know of only one instance of the application of such a design, Fonteyn and Mahall (1981). In both designs, it is useful to include several intermediate densities to reveal possible nonlinearities in the competitive response. These designs, if properly controlled and replicated, should enable one to calculate directly both the per capita effect of one competitor on the other and, since densities are known, the competitive effect of the populations on each other. As will be seen later, few studies have as yet done such a complete analysis.

A third design is sometimes used in which total densities are kept constant. Such experiments have been referred to as "replacement series" by de Wit (1960) or a "reciprocal α" design by DeBenedictis (1974). If used alone it can only reveal the relative, not the absolute, strengths of interspecific versus intraspecific competition. This has the disadvantage that if no significant differences are found in the response to mixed versus single-species treatments at the same total density, it is impossible to decide whether any competition is occurring at all.

Some Practical Problems in Doing Field Experiments

The nature of controls is a crucial factor in doing good field experiments. Because environmental conditions vary in time and space, controls need to be done both contemporaneously with the treatments and as close as possible to them, without at the same time being so close that the process of manipulation itself affects the controls (e.g., removal of individuals may cause immigration from adjacent areas, or cages may change the immediate environment around them [Hurlberg and Oliver 1980; Underwood and Denley 1984]). The aim is to arrange the controls so that, aside from the manipulations of the experiments, the controls and treatments will both experience the same degree of environmental variation. In principle, one can then regard them as differing only in the factor being manipulated experimentally.

Since both populations and environments vary in space and time, however, replication is absolutely essential. In a single experiment, replication of controls and treatments ensures that the results will not be strongly influenced by an extreme event. In addition to such within-experiment replication, repetition of the whole experiment at another place and/or time is also very useful. This ensures that, if competition is occurring but also varying markedly in intensity, the chances of detecting it will be improved.

The second practical problem concerns the population densities to be used in the treatments. Usually one is interested in whether competition occurs over the range of densities found in nature, so this defines the range of the experimental densities. However, it is of little relevance to arrange experimental densities far above the nature range, as some earlier studies did. It is useful to have some treatment density below the range of natural densities because the performance (e.g., fecundity, growth rate, etc.) at the lowest treatment density serves as the baseline against which that at higher densities is compared.

Another problem is that of deciding whether the experimental results can be applied to natural populations. If the treatments consist simply of reducing densities in otherwise natural conditions, there is usually little problem. If, however, mobile animals are placed in enclosures at various densities or if populations are protected from predators, or if the experiment is done in somewhat artificial conditions (e.g., in crops, orchards, artificial ponds, settlement panels placed in conditions different from natural ones, etc.), the problem exists. In such cases, additional information is needed to judge how far the results of such studies can be generalized to natural conditions. Some of the papers reviewed below included such information, indicating that the environmental conditions, individual behavior, and population densities used in the experiment matched reasonably closely those found in populations of the same species living under natural conditions. In cases in which no such information was given, it could probably be gathered by subsequent observations in the communities where the species live naturally.

Finally there is the problem of interpreting the results of a field experiment. Since natural populations interact with many other species, an experimental reduction or increase of a potential competitor population will probably have some effect on species other than the target one. These others, e.g., predators, parasites, prey, competitors, pollinators, etc., may then in turn have a different effect than before on the target species. Such indirect effects should not be considered artifacts; they probably happen commonly when populations fluctuate naturally, and may have important effects on community structure. They make it difficult, however, to be precise about the mechanism that produced the observed result of the experiment.

The best way to deal with this problem is to incorporate it into the design of the experiment. If possible, one should measure the changes in those other species most likely to be affected by the experimental manipulations, and, ideally, manipulate them in turn. The more closely one can observe the changes occurring during the course of an experiment, the greater is the likelihood of discovering the actual mechanism that is operating. This subject is discussed further in the section *Positive and Indirect Interspecific Interactions*.

Measuring the Variability of Interspecies Competition

Probably the most direct way to decide how the occurrence or strength of competition varies is to repeat the field experiment on the same species at a different time or place; this has been done by a few authors. I know of only one instance in which a different person has repeated an experiment (Keough 1983). Ecologists seem reluctant to do this, possibly because they fear that if the result comes out the same it will not be publishable because it is not original enough. A second, indirect, method is to estimate the natural variability of population density or resources by long-term and/or more widespread observations, to see how general (or special) the conditions were during the original experiment. In this way the experiment is set in context, allowing one to judge the conditions under which competition might be expected to occur. If the abundances of the populations or resources can be related to physical variables, it may be possible to use other long-term records, e.g., of weather or hydrography, to estimate indirectly the variability of the populations or resources, as well as how typical the conditions were during the original experiment.

A SURVEY OF FIELD EXPERIMENTS ON COMPETITION

In earlier papers (Connell 1974, 1975) I reviewed many of the field experiments on interspecific competition published through 1973. Since then the number of such field experiments has increased enormously so that my present review can include only a sample. To obtain an objective sample I included only papers from the years 1974 through 1982 in the following six general ecological journals: *Ecology, Ecological Monographs, Journal of Ecology, Journal of Animal Ecology, The American Naturalist,* and *Oecologia*. Although I tried to examine all papers, there were several hundreds to scan, and I may have missed some relevant papers and misinterpreted some results; for any such errors I apologize to the authors. Of all these papers, I have included only those with field experiments satisfying certain criteria. In brief, the study needed to have adequate controls and presentation of data analysis, and sufficient information to judge (*a*) how similar the experimental setup was to natural conditions, and (*b*) whether or not interspecific competition was occurring. The introduction to Appendix A contains details of these criteria.

Among those papers included there still remains a high degree of variability in quality. Degree of replication, rigorousness of design, and thoroughness of data analysis all varied considerably. In some cases the effect of the observed competition on the abundance of the species was clearly demonstrated. In others, such an effect must be inferred from very little data. In the future, as more thorough studies accumulate, inferences from the literature should become more reliable.

For each study I used the following operational definition in deciding whether interspecific competition was affecting a species. Following an experimental change in abundance of a potential competitor, if there was a statistically significant response in the opposite direction in the species being studied, interspecific competition was judged to be occurring. The response could be a change in either breadth of resource use (e.g., food type, microhabitat type) or in abundance (including those variables that affect abundance, e.g., natality, mortality, growth, emigration, immigration, feeding activity, etc.). The response must have been in the opposite direction, e.g., an increase in abundance (or a change in fecundity, mortality, etc., that would increase it) after the competitor was reduced. The degree of statistical significance of the response was judged by comparison to the controls.

The purpose of the present literature survey was to address the questions posed in the introduction: How prevalent and variable is interspecific competition in nature, and what are the relative strengths of intraspecific versus interspecific competition? To answer these questions it was important to get a quantitative estimate of how, for a single species, competition varies in time or space, or in relation to the other species with which it might interact.

To get this information I had to go beyond the methods I had used in previous literature reviews of field experiments (Connell 1974, 1975). Schoener (1983) has recently reviewed the literature on field experiments on competition, with some of these same questions in mind. Like my former reviews, his review is both more extensive and less intensive than the present paper. He has tallied, for 164 studies, the number of species in which competition occurred always, never, or some of the time. In my present survey I have gone a step further. In each paper reviewed, I determined for each species the number of experiments that did or did not show competition. These data provide a quantitative estimate of the variability of competition within species, as well as indicating which estimates are more reliable by virtue of greater sample size. Since it takes more time to extract additional data, I decided to limit my survey to a strictly defined sample of the recent literature, namely the years 1974–1982, in six general ecological journals. This survey yielded 72 papers including 215 species with 527 field experiments that satisfied the criteria described above.

The Frequency of Occurrence of Interspecific Competition

Appendix A gives pertinent information on each of the 72 studies. In this list the frequency of occurrence of interspecific competition is expressed as the fraction: (number of experiments that showed competition)/(total number of experiments done on that species). For the different species, the number of experiments varied. In some cases, competition with several other species was measured, in others, competition with a single other species was measured at several places or in several seasons or years. In contrast, in other cases the response of only one member of a pair of potentially competing species was measured at a single time and place.

To see whether the estimate of the frequency of occurrence of competition was affected by the number of experiments done, the fractions (expressed as percentages) from Appendix A were tallied and analyzed as shown in table 1. As can be seen, the estimate of this frequency is strongly affected by sample size. In studies in which one experiment was done on only one species at a single time and place, competition was found in all but one of the 15 studies, or 93%. In contrast, in those studies in which, as before, only one experiment was done on each species, but more than one species was studied, the frequency fell to 48%. As more experiments were done on each species the frequency dropped even lower (table 1).

This abrupt drop in incidence from 93% when only one species was studied to 48% when two or more species were studied indicates that the former figure is probably an artifact. In many pairs of species studied, one member was being affected by interspecific competition but the other was not (see below). Yet when only one species was studied at a single time and place, in almost every case the investigator apparently chose the member of the pair that was being affected by competition. The survey by Schoener (1983) provides an independent sample of field experimental studies involving only one species at a single time and place. In 29 such studies not included in my present survey, 26 of the 29 species showed competition, or 90%, almost identical to the frequency in my sample.

I suggest that the reason for this trend is to be found in the behavior of scientists, and probably of scientific editors. The original impetus for most scientific studies is the expectation of discovering something. We study competition in a particular context because we hypothesize that it might be occurring there, but if we suspect it is not, we may not even start. Thus a literature survey is already biased toward finding high frequencies of occurrence of competition (or for that matter, any subject searched for). Also, if a single experiment yields a negative result, the investigator may not bother to write it up for publication. If he or she does so, an editor may not accept it; it is probable that authors, editors, and readers are not much interested in single negative results. What this all means is that any estimate of the frequency of interspecific competition gained from a literature survey such as this is probably too high. It is difficult to decide how much of an overestimate the average of 43% for all experiments in table 1 is.

I believe, however, that we can use the results of such a survey to compare different categories of organisms or habitats, unless there is an a priori reason to believe that competition is easier to find with field experiments in certain kinds of organisms or habitats. (For example, wide-ranging animals, the marine plankton, and deeper benthos will probably always be underrepresented in field experiments relative to other groups.) To this end I have tallied the results in three ways: by habitat (terrestrial, freshwater, marine), by broad taxonomic group (plants, invertebrates, vertebrates), and by trophic level (plants, herbivores, carnivores).

Excluding the 15 studies of one species at a single place and time, this leaves 57 papers. In three of these (studies 32, 51, and 67 in Appendix A), no type 1 experiment was done, so that it is uncertain whether interspecific competition was occurring. Of the remaining 54 papers having type 1 experiments, some interspecific competition was found in 83%; Schoener (1983) found this percentage to be 90% in the 164 studies in his survey. This method yields an overestimate,

TABLE 1

EFFECT OF SAMPLE SIZE ON ESTIMATES OF THE FREQUENCY OF OCCURRENCE OF INTERSPECIFIC COMPETITION

No. of Experiments per Species Studied	Studies of One or More Species at a Single Time and Place		Studies of the Same Pair of Species at Different Times or Places		All Studies			
	No. of Species	Mean, Over All Spp., of % of Exp. per Sp. Showing Competition	No. of Species	Mean, Over All Spp., of % of Exp. per Sp. Showing Competition	No. of Species	No. of Exp.	% of Exp. per Sp. Showing Competition	
							Mean	Coeff. Var.
1 (only 1 sp. in the study)	15	93	...	...	15	15	93	28
1 (>1 sp. in the study)	65	48	...	...	65	65	48	106
2	14	64	49	46	63	126	50	92
3	22	27	15	49	37	111	36	112
4	10	⎫	8	⎫	18	72	38	106
5	3	⎪		⎪	3	15		
6	1	⎬ 22	1	⎬ 38	2	12		
7	3	⎪		⎪	3	21		
8	4	⎭		⎭	4	32	⎫ 9	⎫ 110
11	2		1		3	33	⎬	⎬
12			1		1	12	⎪	⎪
13			1		1	13	⎭	⎭
Total	124*	41.0*	76	44.8	215	527	43.0*	105*

NOTE.—For each species, the percentage of experiments showing competition for one response was used as a datum.
* These totals exclude the first line (studies of only 1 sp. at a single time and place).

however, since even in studies with positive results some of the species showed no competition. In 55% of the 200 species in the 54 papers some interspecific competition was found; Schoener (1983) found it in 76% of the species in his survey. This method also yields an overestimate since in those species where some interspecific competition was found, it often did not occur in all places or times, or against all other species in the study. To take this variation into account, I used, for each species, the percentage of experiments showing competition; these are the same percentages tallied in table 1. When these percentages were averaged (excluding the above 15 studies), competition was found 43% of the time in the 200 species (table 1). (In this and subsequent analyses, the experimental result for only one response variable per species was included. Since in some studies several variables were measured, e.g., changes in density, mortality, fecundity, growth, etc., a decision had to be made as to which to use. I decided to use the response variable most directly related to a change in abundance. This is the first response listed in each study in Appendix A.)

In tables 2 and 3, the average over all species of the fractions for each species is given for each category of habitat/organism. This weights species equally but experiments unequally because different species had different numbers of experiments. So I also give the percentage of all experiments that showed competition in each category, which weights experiments equally. Since in these analyses I was interested in comparing populations open to the influence of grazers, predators, vagaries of weather, etc., it was necessary to exclude the 13 studies done in enclosures. These were included in later analyses. The studies involving only one species done at a single time and place were also excluded on the basis of the discussion above.

The comparisons in these two tables can be used to test some a priori hypotheses that have been proposed concerning the incidence of interspecific competition. Hairston et al. (1960) (hereafter HSS) proposed that in terrestrial habitats herbivores as a trophic level are not likely to compete for common resources, being controlled by their predators, whereas the other trophic levels of producers (plants), carnivores, and decomposers should compete. If we assume that individual populations represent the behavior of their trophic level, this modification of the original hypothesis can be tested with the present data, with one constraint.

Slobodkin et al. (1967) specified that the hypothesis in HSS applies only to those terrestrial herbivores that feed on the plant itself, not on its products such as nectar, pollen, and seeds. Nine of the species in three studies (nos. 56, 57, and 61) satisfy this criterion and are listed separately, as phytophages, in table 2. Their average frequency of competition did not differ significantly from either the terrestrial plants or carnivores. Slobodkin et al. (1967) also predicted that those herbivores that eat nectar, pollen, or seeds would compete, in contrast to the phytophagous herbivores. The average frequency of competition of the 10 species in the former category were tested against that of the nine phytophagous species and found not to differ. None of these results are consistent with the HSS hypothesis. However, since the sample sizes of terrestrial phytophagous herbivores are small and all were insects living on *Heliconia* plants, it is hardly a definitive test of the hypothesis.

TABLE 2

Literature Survey of Field Experiments on Interspecific Competition, Comparing Three Trophic Level Groups in Three Habitats

A. Average of percentage of experiments done for each species that showed interspecific competition

	Terrestrial		Marine		Freshwater		Total	
	No. Spp.	Ave. %	No. Spp.	Ave. %	No. Spp.	Ave. %	No. Spp.	Ave. %
Plants	50	39	22	59	2	50	74	45
Herbivores	19	21	9	78	0	...	28	39
(Phytophages)	(9)	(18)						
Carnivores	17	10	4	75	1	67	22	24
Total	86	29	35	66	3	56	124	41

B. Percentage of all experiments in each category that showed interspecific competition

	Terrestrial		Marine		Freshwater		Total	
	No. Exp.	%	No. Exp.	%	No. Exp.	%	No. Exp.	%
Plants	205	30	31	68	2	50	238	35
Herbivores	45	20	13	69	0	...	58	31
(Phytophages)	(22)	(23)						
Carnivores	36	11	5	60	3	67	44	20
Total	286	26	49	67	5	60	340	32

Note.—Student's t tests on the percentages for each species used in Part A above to test certain a priori hypotheses (see text). Phytophagous herbivores were not significantly different ($P > .05$) from either plants or carnivores in the terrestrial habitat ($t = 1.40, .75$, respectively). This is also true for the same comparisons using all herbivores, in terrestrial ($t = 1.66, 1.04$), and marine ($t = .99, .09$), and total ($t = .54, 1.18$). Plants were different from carnivores in the terrestrial habitat ($t = 2.64, P < .02$) but not in marine or total ($t = .59, 1.90$). Between terrestrial vs. marine habitats, plants were not different ($t = 1.78$), but herbivores and carnivores were different ($t = 3.62, P < .002; t = 3.32, P < .01$). Within terrestrial herbivores, the 9 phytophages were not different ($t = 0.29$) from the 10 species that ate plant products such as nectar, pollen, or seeds.

This table does not include studies of one species at a single place and time (study nos. 1, 2, 3, 6, 13, 29, 30, 38, 41, 43, 44, 48, 49, 50, 53), nor studies done in enclosures (study nos. 7, 9, 11, 35, 36, 37, 39, 45, 46, 59, 66, 72), nor studies of omnivores (study nos. 5, 8, 12, 28, 42, 57 (3 spp.), 58 (2 spp.), 62, 71).

Schoener (1983) used the numbers of studies or of species showing some competition to test the HSS hypothesis. For terrestrial populations the sample size of unenclosed populations of phytophagous herbivores was, like mine, very small (5 studies, 8 species). One study with three species of insects overlapped with my sample; the others in Schoener's sample were four rodent species and one insect. Comparisons with other trophic categories showed a lower frequency of competition in the eight phytophagous species as compared to producers and carnivores, in two out of three statistical tests on unenclosed populations. Although Schoener (1983) considers that the results support HSS, the small sample size of unenclosed terrestrial studies in Schoener's (1983) survey, as in my own, makes any judgment quite tentative at this stage.

TABLE 3

LITERATURE SURVEY OF FIELD EXPERIMENTS ON INTERSPECIFIC COMPETITION, COMPARING THREE
TAXONOMIC GROUPS IN THREE HABITATS

A. Average of fraction of experiments done for each species that showed interspecific competition

	TERRESTRIAL		MARINE		FRESHWATER		TOTAL	
	No. Spp.	Ave. %	No. Spp.	Ave. %	No. Spp.	Ave. %	No. Spp.	Ave. %
Plants	50	39	22	59	2	50	74	45
Invertebrates	23	17	24	31	0	...	47	24
Vertebrates	25	23	9	89	1	67	35	41
Total	98	30	55	52	3	56	156	38

B. Percentage of all experiments in each category that showed interspecific competition

	TERRESTRIAL		MARINE		FRESHWATER		TOTAL	
	No. Exp.	%	No. Exp.	%	No. Exp.	%	No. Exp.	%
Plants	205	30	31	68	2	50	238	35
Invertebrates	57	16	37	32	0	...	94	22
Vertebrates	47	23	10	90	3	67	60	37
Total	309	26	78	54	5	60	392	32

NOTE.—Student's t-tests as in table 2. Invertebrates had a lower percentage than plants in terrestrial ($t = 2.18$, $P < .05$), marine ($t = 2.09$, $P < .05$), and total ($t = 2.58$, $P < .02$). Invertebrates were also lower than vertebrates in marine ($t = 3.65$, $P < .001$), but not in terrestrial or total ($t = 0.49$, 1.72). The studies done in enclosures and those done with only one species at a single place and time were excluded, but omnivores were included in this table (see table 2).

Although the original HSS hypothesis was limited strictly to terrestrial phytophagous herbivores, it is interesting to see whether the general concept might apply more broadly. Since the comparisons of herbivores as a group to either plants or carnivores in all habitats in table 2 show no significant differences, they do not support such an extension of the hypothesis.

Schoener (1983) also compared trophic levels in other habitats; the sample sizes of herbivores in unenclosed experiments in marine and freshwater habitats (19 and 12 species, respectively) were somewhat larger than in the terrestrial comparison discussed above. The tests with marine herbivores were never consistent with HSS, while the freshwater ones were so in two thirds of the tests. When number of studies showing some competition was used by Schoener (1983), rather than number of species, marine and freshwater studies showed no significant differences between trophic levels. Terrestrial studies are consistent with HSS but the sample size of unenclosed herbivores is small, only five studies. Thus his evidence is not consistently supportive, and does not take into account the variation within species, as this paper does. When both his and the present survey are taken together the evidence is still insufficient and too inconsistent either to support or reject HSS.

A second hypothesis concerning trends in competition across trophic levels is that of Menge and Sutherland (1976). They predict that, in the organization of guilds within trophic levels, interspecific competition will be more prevalent at higher levels. This hypothesis predicts the same trend as HSS for the upper two trophic levels: Predators will compete more than herbivores. For the lower two levels it predicts the opposite of HSS: Herbivores will compete more than plants. The data in table 2 support neither of these predictions; only one of the comparisons between trophic levels within habitats showed a significant difference, and it was in the direction opposite to that predicted. This analysis is perhaps not a strong test, since this hypothesis also implies that populations at a lower trophic level should compete if they have somehow escaped from control by their predators.

A third a priori hypothesis which could be tested with these data is the prediction that interspecific competition is less likely to occur in habitats or groups of organisms that are more vulnerable either to harsh physical conditions or to natural enemies (Connell 1975). Smaller-bodied individuals are often more vulnerable than larger ones to both these hazards. In my paper (Connell 1975, figs. 2–5) I suggested a model of how body size might interact with gradients in the physical environment and predation to affect the degree of mortality from these agents and so the degree of interspecific competition experienced by an organism.

The invertebrates in table 3 all have smaller body sizes than either the plants or the vertebrates. (There were no small or microscopic plants in the studies reviewed here.) As a consequence, my hypothesis would predict that the invertebrates would be less likely to compete than would plants or vertebrates. The analyses in table 3 are consistent with this prediction for marine but not for terrestrial species. Also, since the terrestrial herbivores (insects and small mammals) are probably more vulnerable to predation than the marine herbivores in these studies (sea urchins), the same predictions apply. The analysis in table 2 confirms this prediction. Schoener (1974) had also suggested that smaller animals would compete less than larger ones, and analysis of the data from his recent survey (Schoener 1983) is consistent with this prediction.

One other clear trend shown in tables 2 and 3 is that, without exception, terrestrial organisms showed a lower frequency of interspecific competition than marine ones. With regard to the plants, I suspect that the lower frequency of terrestrial than marine competition may be a consequence of the fact that most of the terrestrial studies were made of herbs on grazed pastures, mown turfs, or on sites of recently abandoned agriculture. Grazing and mowing act in the same fashion as predation in reducing competition (Harper 1969; Connell 1975); also, on recently abandoned fields the abundance may not have reached levels high enough to result in strong interference between the plants. In contrast, the marine plant examples all come from natural stands of larger algae which may not have been greatly affected by grazing. This is not always the case; other grazing studies have indicated that sea urchins are capable of greatly reducing the abundance of larger algae (Paine and Vadas 1969).

The terrestrial invertebrates (insects and spiders) were about the same size as the marine invertebrates (except for sea urchins), so that vulnerability to predation probably cannot account for the differences. The same applies to the vertebrates; in any case the studies of marine vertebrates were too few to permit useful comparisons. Probably there is no single explanation for the contrast in competition between terrestrial and marine organisms.

The results to this point all refer to responses that were either changes in abundance or in a variable that could affect abundance. A few studies measured changes in resource use, i.e., competitive niche release. These 14 papers indicate that competitive release occurred in about three quarters of these species (see table 4). Five studies involved only one experiment at a single time and place, and as indicated in table 1, a tally of such estimates tends greatly to overestimate the frequency of occurrence of interspecific competition. One paper (Holbrook 1979; study no. 28) studied competitive release of a species paired with more than one other species; none was done with the same pair at different times or places.

Variation in Occurrence of Interspecific Competition

A species may vary in at least two ways in the degree to which it competes. It may compete with some species in the community but not others, and it may compete with a particular other species at some times or places but not at others. The mechanisms producing the former sort of variation are difficult to investigate, since they are in part a reflection of the evolution of the species. The mechanisms producing variation in space and time are more amenable to investigation, since they may sometimes be correlated with contemporary environmental variations.

Thirty-six of the 72 studies gave data on interspecific competition between the same species at different places or times. The results are shown in table 5. In about half of the species in these studies no interspecific competition was found, so no conclusions can be drawn concerning variation. In those species that showed some competition, however, 59% showed annual variation and 31% showed spatial variation. Where data on annual variation are available, competition occurred in 35% of the years; the corresponding figure for spatial variation was 46% of the sites studied.

Schoener (1983) also discusses temporal variability in interspecific competition. Eleven of the studies in his review showed annual variability, three others possibly did so, and 12 showed none. Within-year variability occurred in one case and not in two others. Thus in the studies that had evidence for it, competition was variable in about half of the studies in his review. This is about the same degree of variation that emerges from my analysis of variability within species. Apparently it is not rare for competition to vary in either time or space, although Schoener (1983) and I differ in this judgment.

Reasons for these variations have been suggested in some cases. Wiens (1977) has proposed that competition might occur only when resources are scarce. I will discuss three studies of lizards in detail since they provide very interesting evidence concerning this hypothesis. Dunham (1980, study 15) studied *Urosaurus ornatus* and *Sceloporus merriami* in 1974–1977 in Texas, and Smith (1981, study 58) studied the same *Urosaurus* sp. and *Sceloporus virgatus* in 1973–1976 in Arizona. Insect prey became scarce in dry years; two of the four years (1975 and

TABLE 4

Field Experiments on Competitive Release in Niche Breadth (Response measured was a change in horizontal or vertical distribution toward that of a competitor, following an experimental reduction in the abundance of that competitor.)

Study No.	Habitat	Organism	No. of Taxa	Change Measured	No. of Exp. per Taxon (no. comp. found/total no.)
38	freshwater	plant	1	horizontal zonation from pond edge	1/1
34	marine	algae	2	intertidal ht., 2 sites	2/2, 2/2
70	terrestrial	spiders	2	ht. of web	0/1, 0/1
6	marine	snails	1	intertidal ht.	1/1
50	marine	snails	1	intertidal ht.	1/1
5	marine	crabs	2	intertidal ht.	1/1, 0/1
27	marine	fish	2	subtidal depth (2 yr for 1 sp)	0/1, 2/2
33	marine	fish	2	subtidal depth	1/1, 1/1
54	marine	fish	5	horizontal distribution	5@ 1/1
64	terrestrial	lizard	1	aboreal vs. ground	0/1
8	terrestrial	mammal	2	altitude on mt.	0/1, 1/1
28	terrestrial	mammal	3	aboreal vs. ground	0/1, 1/1, 2/2
28	terrestrial	mammal	3 (same)	horizontal, veg. type	0/1, 1/1, 1/2
42	terrestrial	mammal	2	horizontal, veg. cover	1/1, 1/1
49	terrestrial	mammal	1	horizontal, veg. type	1/1
55	terrestrial	mammal	2	horizontal, veg. type	0/1, 0/1

Summary: 11 of the 15 studies (73%) showed some interspecific competition; 22 of the 29 species (76%) showed some interspecific competition; 26 of the 37 experiments (70%) found interspecific competition.

TABLE 5

Variation in Time and Space in the Occurrence of Interspecific Competition (Entries are the number of species that were observed for the number of years, seasons or sites shown, and that competed in the given % of those times or places.)

	No. of Spp. with Following % of Yr or Sites that Showed Competition					Species Showing Some Competition		All Species	
	0	1–33	34–66	67–99	100	No.	% Spp. with Some Variation	No.	Ave. % Yr or Sites with Comp.
A. Annual Variation (seasonal in parentheses)									
No. Yr. Obs.									
2	12 (5)		7 (3)		6 (6)	13 (9)	54 (33)	25 (14)	38 (54)
3	7			3	1 (1)	4 (1)	75 (0)	11 (1)	27 (100)
Total ...	19 (5)		7 (3)	3	7 (7)	17 (10)	59 (30)	36 (15)	35 (57)
B. Spatial Variation									
No. Sites Obs.									
2	18		2		15	17	12	35	46
3	5	1			3	4		9	
4	1	1	3	1	2	7	58	8	45
13		1				1		1	
Total ...	24	3	5	1	20	29	31	53	46

Note.—A. Study nos. with data on annual var.: 11, 12, 15, 16, 22, 27, 29, 45, 46, 52, 58, 62, 64, 68, 71; seasonal var.: 9, 17, 35, 46, 47, 59, 66. B. Study nos. with data on spatial var.: 4, 7, 9, 10, 12, 19, 20, 23, 24, 25, 34, 39, 40, 46, 56, 68, 72.

1977) were exceptionally dry in Texas, one of the four years (1974) was exceptionally dry in Arizona. Wiens' (1977) hypothesis predicts that competition would be more likely to occur in the drier years when food resources are scarcer. Evidence consistent with this hypothesis was found as follows. In *Urosaurus* in Texas, competitive effects were found on growth in two dry years but not in two wet years, and on foraging success in early summer when food was scarce but not in late summer when food was abundant. For *Urosaurus* in Arizona, competitive effects on survival of young females were found in one dry year, not in two wet years. In *Sceloporus virgatus* in Arizona, competitive effects on growth and survival of young females were found in the dry year but not in the next wet year. However, some evidence in the Texas study is inconsistent with the hypothesis. Competitive effects on yearling survival of *Urosaurus* occurred in one wet year but not in a dry year. Also, competitive responses in density of *Urosaurus* were found in both a wet and a dry year, but not in the other dry year. In a third study of effects on *Sceloporus undulatus* of removal of two other lizard species in another location in Arizona, Tinkle (1982, study 64) found no competitive effects in four different responses in two years. Overall, as shown in Appendix A, interspecific competition occurred rarely in all three studies; when it was found, some responses were consistent with Wiens' (1977) hypothesis, others were not. *Urosaurus ornatus* showed more competitive effects in the drier Texas site than in the more mesic Arizona site, which is consistent with Wiens' hypothesis. These excellent studies demonstrate the value of measuring different responses over long periods.

Another study showing annual variation is that of Ericksson (1979, study no. 16) of carnivorous ducks competing with fish in Swedish lakes. In two years there were significantly more ducks on lakes without fish than on those with fish added; in the third year the difference was not statistically significant. In the year when the numbers of ducks were most different between treatments the insect prey were also the most different. However, in the other two years the competitive effects were not correlated with abundance of food. Overall these results are consistent with Wiens' (1977) hypothesis.

One study of marine invertebrates gave evidence of annual variation. Sutherland (1978, study no. 62) removed two sessile colonial species from fouling panels in each of two years. The effects of these treatments differed in the two years, because all species, both those that were experimentally manipulated and those being affected, varied greatly in abundance between the two years. These differences may have been the result of temporal variations in the rates of recruitment from the plankton of many of these species.

Spatial variation was observed in several instances. Miles (1974, study no. 40) studied the effects of removing the canopy and litter on eight species of plants on four forest sites with different soils. In one species no competition was found on any site; in two species competitive effects were found on all four sites, while competition varied between sites in the other five species. The strongest effect was caused by interference with germination by the litter from the canopy plants. Differences in competition between sites may have resulted from variations in the soils, or in the type and amount of litter present. One other terrestrial study had data on spatial variation. Seifert and Seifert (1976, study no. 56) found competition between a pair of insects in the flower bracts of one species of host plant but not in a congener. As shown in Appendix A, interspecific competition was rare in this ephemeral community and this variation between hosts may be simply a chance occurrence.

Spatial variations in competition in marine algae were found by Dayton (1975, study no. 10). Two species showed no spatial variation; the third, a group of "fugitive" species, showed no variation in competition with *Lessoniopsis* but did show variation in competition with *Hedophyllum*. The reason for the latter variation was simply that *Hedophyllum* was relatively rare in one of the sites so that its removal there did not open up much of the free space which the fugitive species require. Menge (1976, study no. 39) found that barnacles were outcompeted by mussels in only four of the 13 sites in which both species were capable of covering all the space. Three of the four were in places where predatory snails could not feed, either because of heavy waves or a high shore refuge. The fourth was classified as a temporary escape from predation. In the other nine sites, heavy predation prevented the mussels from outcompeting the barnacles.

In summary, interspecific competition may not occur for several reasons. In the cases of the sessile marine animals studied by Sutherland (1978), recruitment may sometimes fail to occur. In others, e.g., Dayton (1975), the species being experimentally reduced may be so rare that effects on other species are too slight to be detectable against the background of normal variation. Third, in some years or sites resources may not be limiting (Wiens 1977); some evidence of this comes from the studies of lizards and ducks described earlier. Fourth, predation may reduce the populations to such low levels that they are not pressing on their resources. Some examples from this survey are: predation by snails on mussels (Menge 1976); by sea urchins on algae (Dayton 1975; Duggins 1980); and by various predators on tadpoles (DeBenedictis 1974). Lastly, the process of competition may be interrupted by physical disturbances, as seen in marine intertidal plants and sessile animals (Sousa 1979; Taylor and Littler 1982).

The Relative Strengths of Interspecific Versus Intraspecific Competition

If we are interested in the coexistence of competing species, it is not sufficient to measure only interspecific interactions. As discussed earlier, it is necessary to do experiments designed particularly to distinguish the two sorts of competition. Of the 72 papers, 14 did this, studying 42 species with 123 experiments. As shown in table 6, for 31% of the species and 17% of the experiments interspecific competition was stronger than intraspecific competition. Eleven of the 14 studies also performed type 1 experiments on the same species and found interspecific competition for 71% of the species and 40% of the experiments. Thus a comparison of the results of the two types of experiments shows that when interspecific competition was found by a type 1 experiment, a type 2 experiment showed it was stronger than intraspecific in less than one-quarter of those cases.

It is important to evaluate how representative this sample is. As shown in table 6, all three habitats are represented, as well as the three taxonomic and trophic groups used earlier. The terrestrial habitat and invertebrate herbivores are overrepresented, however. Also, the percentage of species showing interspecific competition with type 1 experiments in table 6 (71%) is higher than the average of 55% for all species; however, the 40% of experiments showing competition is only slightly higher than that in tables 2 and 3.

Asymmetry and the Rank Order of Competitive Ability

Competition is the term used when each species of a pair has a negative effect on the other. If only one species of a pair is affected negatively, this has been termed "amensalism" (Burkholder [1952] used the signs $(-,-)$ for competition, $(0,-)$ for amensalism). In the present survey, both members of a pair of species were studied with type 1 experiments in 98 cases. Of these, 44 pairs showed no interaction, 33 pairs showed apparent amensalism with only one species affected, and 21 pairs showed competition with a reciprocal negative effect (table 7).

If one member of a pair of species that are competing for the same resources is strongly affected by the second, it seems reasonable to expect that there could be some reciprocal effect, however slight, but with the weaker one being undetect-

98

TABLE 6

RESULTS OF ALL STUDIES WITH TYPE 2 EXPERIMENTS MEASURING BOTH INTRASPECIFIC AND INTERSPECIFIC COMPETITION, COMPARED TO THE RESULTS OF TYPE 1 EXPERIMENTS FROM THE SAME PAPERS (Entries are number of experiments and, in parentheses, number of species.)

				No. of Experiments (no. spp.)					
				Type 1 Experiment		Type 2 Experiment			
				Intersp. Comp.					
Study No.	Habitat/ Organism & Trophic Level	No. of Taxa	Response	Found	Not Found	Inter > Intra	Inter = Intra	Intra > Inter	No. Comp. Found
17	ter./plant	2	physiol.	4 (2)		3 (2)	1	4 (2)	
32	ter./plant	4	wt.			4 (2)	4	1	
51	ter./plant	2	growth			2 (1)	3 (1)		
37	ter./inv./herb.	7	repr.	16 (7)			7 (3)	9 (4)	
56	ter./inv./herb.	6	dens.	5 (4)	19 (2)	4 (3)		12 (3)	8
57	ter./inv./herb.	4	dens.	1 (1)	11 (3)			3 (1)	9 (3)
70	ter./inv./herb.	2	surv., repr.		2 (2)			2 (2)	
7	fw./inv./herb.	2	repr.	6 (2)		3 (1)	1	2 (1)	
59	fw./inv./herb.	2	pop. gr.	8 (2)	8	1 (1)		1 (1)	
67	fw./vert./carn.	2	surv.				2 (2)		
9	mar./inv./herb.	3	mort.	3 (2)	5 (1)	3 (2)	2	3 (1)	15
46	mar./inv./herb.	2	growth	1 (1)	22 (1)			8 (2)	
66	mar./inv./herb.	3	mort.	3 (2)	5 (1)		2	2 (2)	
54	mar./vert./herb.	1	feeding	1 (1)		1 (1)		1 (1)	
Total no. species				24	10	13	6	20	3
% of total				70.6		31.0	14.3	47.6	7.1
Total no. experiments				48	72	21	22	48	32
% of total				40.0		17.1	17.9	39.0	26.0
% of exp. that found competition						23.1		76.9	

NOTE.—ter. = terrestrial, inv. = invertebrate, herb. = herbivore, fw. = freshwater, carn. = carnivore, mar. = marine, vert. = vertebrate, physiol. = physiological, wt. = weight, repr. = reproduction, dens. = density, surv. = survival, pop. gr. = population growth, mort. = mortality.

TABLE 7

Asymmetry in Interspecific Competition; Evidence from All Studies in which Type 1 Experiments Were Performed on Both Members of a Pair of Species (If more than one pair was studied or more than one response variable per pair, this information is indicated in parentheses after the study code number.)

1. SYMMETRICAL COMPETITION (both members of the pair responded)
Study nos: 7 (fecundity only), 9 (growth 1 pr), 15 (first yr survival), 17, 29, 33, 35, 36, 37 (8 pr), 42 (4 variables), 58 (survival only), 59 (2 pr), 69. Total 21 pr.

2. ASYMMETRICAL COMPETITION (only one member responded)
Study nos: 4 (5 pr), 5, 7 (growth only), 8, 9 (2 pr), 10 (2 pr), 15 (6 var.), 18 (4 pr), 21, 24, 27, 28 (2 variables), 35 (2 pr), 42 (2 variables), 46, 56 (5 pr), 57, 58 (growth young females), 59, 62, 66 (3 pr). Total 33 pr exclusive of those in symmetrical category.

3. NO COMPETITION (shown by either member of the pair)
Study nos: 4 (17 pr), 9 (mortality, 1 pr), 10, 11, 18 (9 pr), 28 (1 variable), 42 (1 variable), 45, 55, 56, (6 pr), 57 (5 pr), 58 (several variables), 60, 61, 70. Total 44 pr, exclusive of the above two categories.

4. REVERSALS OF RANK ORDER: evidence for reversals in the order of competitive superiority in a pair is presented only in the following studies: (*a*) reversals occurred in study nos. 15, 35, 46 (see text); (*b*) no reversals occurred in study nos. 4, 7, 9, 24, 56, 58, 59, 66.

---

able against the background environmental variation. Therefore I suggest that the term asymmetrical competition be used, as Lawton and Hassell (1981) have done, in place of amensalism. The latter term has never taken hold in ecology; all the authors in the present survey used competition, never amensalism.

Of the 54 pairs in table 7 that showed some competition, 61% were strongly asymmetrical. Lawton and Hassell (1981), in a survey of insect species, found 66% of 35 pairs to be strongly asymmetrical; 15 of these pairs are included in the present survey. In Schoener's (1983) survey, 61 studies contained experiments on both members of one or more pairs (32 of these studies are included in the present survey). Eighty-four percent of these 61 studies had some pairs showing strong asymmetry. Clearly it is the rule rather than the exception.

With strong asymmetry one can easily rank the species in competitive ability. In an asymmetrical pair, the species unaffected in a field experiment should be superior in competition to the one affected. Even in symmetrical pairs, exact equivalence is seldom expected, although it may be more difficult to rank the species than when strong asymmetry exists. If the rank order remains the same, and if interspecific competition is stronger than intraspecific in the superior species, in theory the inferior one should eventually be eliminated, all other things being equal (which they seldom are).

However if the rank order of competitive superiority were sometimes reversed, competitive elimination is less likely. Few studies are sufficiently comprehensive to provide evidence concerning such reversals. To obtain such evidence requires at least two experiments on both members of a pair, which is rarely done. In the present survey, 11 studies contained sufficient data to judge this (see table 7); in eight of these no reversals were found. In the other three the order of competitive superiority changed as follows. Lynch (1978, study 35) found that *Ceriodaphnia* was superior in a summer experiment whereas *Daphnia* was superior in an autumn experiment. Dunham (1980, study 15) found that when using first-year survival as the measured response, only *Urosaurus* was affected in 1974, whereas only *Sceloporus* was affected in 1975. Using other responses, only *Sceloporus* was affected for adult survival, whereas only *Urosaurus* was affected for all other responses. Peterson (1982, study 46) found that, when using either growth or gonad mass as a response variable, *Protothaca* was affected but not *Chione,* but when recruitment was measured, only *Chione* was affected. Thus the rank order of competitive superiority within a pair of species changed in these three studies, either in different seasons or years or among the different response variables measured.

Field experimental evidence on reversals in the rank order of competitive superiority between members of a single pair of species is sparse, as shown by the present survey. However, another source of evidence exists in the studies of interactions between sessile marine organisms on hard substrates. Direct observations of contacts between neighbors can yield evidence as to whether one is killing its neighbor by overgrowth, allelopathy, or other means of aggressive interference. If enough observations are made of contacts between individuals or colonies it is possible to test the null hypothesis of equivalence in wins between two species (Kay and Keough 1981). A recent review (Connell and Keough 1983) indicates that there is now sufficient evidence to test this hypothesis in a few studies (Jackson 1979; Buss 1980; Kay and Keough 1981; Russ 1982; Rubin 1982; Connell and Keough 1983). In some pairs of species, competition was asymmetrical and consistent, one species always winning. In other pairs there were frequent reversals, each species winning some of the contests. In other contests, neither species won; some of these standoffs lasted many years (Connell 1976).

What determines competitive superiority? Among many possibilities, size has been suggested as being important by several authors, the larger species usually (but not always) being superior (see review in Schoener 1983). With organisms that compete for space, and in particular the clonal ones such as plants and attached colonial animals, competition occurs mainly between neighbors. Here, individual size varies enormously and usually the larger neighbor is superior, regardless of species; taller forest trees suppress smaller ones, larger colonial animals win over smaller ones (Buss 1980; Russ 1982). Thus competitive ability, rather than being species specific becomes size specific or age specific. In such cases, reversals of competitive rank between species should be common, reducing the degree of asymmetry.

Positive and Indirect Interspecific Interactions

Some responses to the interspecific experimental manipulations in the present survey were positive. These occurred in four studies of plants (nos. 4, 10, 18, 63 in Appendix A), seven studies of herbivores (nos. 14, 35, 37, 46, 56, 57, 61), and three studies of omnivores or carnivores (nos. 15, 28, 62). Most of these positive responses involved only one member of a pair; mutually positive responses occurred only in studies 37 and 56. In some studies (nos. 4, 15, 18, 46) a large number of experiments were done, so that some of the results in these studies may be attributable to chance.

What mechanisms might account for these positive interactions between species on the same trophic level? Direct positive interactions between species are clearly possible. For example, in study no. 10 (Dayton 1975) the larger overstory algae probably protected the smaller understory species from desiccation or radiation damage. The larger species of sea urchin in study no. 14 (Duggins 1981) apparently protected the smaller species from predatory starfish. Alternatively, negative effects acting indirectly through other species may produce positive outcomes. For example, let us assume that species B competes with both species A and C, but the latter two do not compete. If A were experimentally increased, this could cause B to decrease, which would allow C, being released from competition, to rise. A reasonable but incorrect interpretation is that A and C interact positively. This possibility was suggested by Fowler (1981) to account for the positive responses in study no. 18; other examples of such complex competitive interactions have been studied and discussed by several other authors (Neill 1974; Wilbur 1972; Smith-Gill and Gill 1978).

A second type of indirect interaction involves predators that are specialized on particular prey species. For example, if the prey compete but the predators do not, an experimental increase in one predator might cause a decrease in its prey A, a consequent increase in prey B because it was released from competition, and so an increase in its predator. A reasonable but incorrect interpretation of this experimental result is that the predators were interacting positively. This idea was called "complementary feeding niches" by Dodson (1970) and has been discussed theoretically by Levine (1976) and Vandermeer (1980). A mechanism similar to this was suggested by Lynch (1978) to explain the positive responses in his study.

The converse is also possible. If prey A and B were interacting positively instead of competing in the above experiment, both would decrease, causing a reduction in the second predator. An investigator ignorant of the prey interactions might incorrectly conclude that the predators were competing.

Quinn and Dunham (1983) point out the problem posed by such multiple causes acting simultaneously, and warn against testing "univariate hypotheses" that postulate, for example, that only competition or only predation is responsible for a given pattern. This problem has bothered ecologists for many years and to overcome it several approaches have been tried. First and foremost is the necessity of studying the system from all angles so as not to miss the important interactions; there are no quick and dirty ways to do ecology. Second is the testing of several hypotheses; Jackson (1981) has pointed out that plant ecologists have been doing field experiments on the effect of grazers on plant competition since at least 1917. Although it has not often been done, the technique of simultaneously varying the densities of both competitors and predators in a nested experimental design is to be recommended. For example, by doing this I was able to indicate the effect of predation on interspecific competition in intertidal animals (Connell 1961), a situation very similar to the example given by Quinn and Dunham (1983). Two studies from the present survey performed simultaneous manipulations of competitors and predators (DeBenedictis 1974; Peterson 1982), while Dayton (1975) and Sousa (1979) studied both grazing and competition in separate experiments. As Simberloff (1983) suggests, the fact of multiple causality in ecology only demands more ingenuity in framing hypotheses and choosing tests whose outcomes are unambiguous.

SUMMARY

In a strictly defined sample of competition studies using controlled field experiments, covering 215 species and 527 experiments, competition was found in most of the studies, in somewhat more than half of the species, and in about two-fifths of the experiments. In most of these experiments interspecific competition was not distinguished from intraspecific competition. In the few studies in which the two were separated, interspecific competition was the stronger form in about one-sixth of all experiments done. When competition was demonstrated, intraspecific competition was as strong or stronger than interspecific in three-quarters of the experiments.

Some evidence from this literature survey suggests that negative results may be underrepresented, so that the absolute values of these figures may be too high. Since this bias should apply also to studies of all taxa, habitats, or other interactions it should not greatly affect estimates of the relative prevalence of competition. Since these estimates come from field experiments open to other influences such as predators, grazers, weather, disturbances, etc., they should provide a fair approximation of the relative prevalence of interspecific and intraspecific competition in natural ecological communities.

The prevalence of competition in these studies varied. Marine organisms showed consistently higher frequencies of competition than terrestrial ones as did large-sized organisms as compared to smaller ones. Plants, herbivores, and carnivores showed similar frequencies of competition in all habitats compared. The incidence of competition varied considerably from year to year and place to place.

In some categories, evidence concerning competition is sparse. More studies are needed of all freshwater species, marine vertebrates, parasites, effects on resource partitioning, and particularly the relative strengths of interspecific versus intraspecific competition. When both members of a pair were studied and some competition found, only one member was affected in well over half the experiments. Such strong asymmetrical competition is not always consistent in direction; reversals in the rank order of competitive superiority have been demonstrated by field experiments and direct observations.

Some positive interactions were found. These may have been a consequence of actual positive influences or of negative ones acting indirectly through other species. The latter may also apply to some of the negative interactions interpreted as competition in these studies. If only the input and output of an experiment are known, it is difficult to decide what mechanism produced the observed effect. While many of the experiments probably have been correctly interpreted, the present survey illustrates how difficult it is to produce a clear and unambiguous demonstration of interspecific competition.

ACKNOWLEDGMENTS

I would like to thank the following for their discussions and criticisms of this paper: J. Bence, R. Black, A. Blaustein, J. Choat, S. Cooper, P. Dayton, R. Dean, E. Denley, J. Dixon, B. Downes, R. Doyle, M. Dungan, A. Dunham, P. Fairweather, M. Fawcett, N. Hairston, S. Holbrook, K. Hopper, J. Jackson, M. Keough, B. Mahall, M. Marsh, W. Murdoch, C. Osenberg, R. Paine, D. Peart, C. Peterson, M. Price, J. Quinn, P. Raimondi, M. Saunders, R. Schmitt, S. Schroeter, B. Sheehan, D. Shiel, D. Smith, A. Smythe, W. Sousa, D. Spiller, A. Stewart-Oaten, S. Swarbrick, A. Underwood, J. Wild, B. Menge, and T. Schoener.

APPENDIX A

Summary of Experimental Results

Entries for each study are: study code no., author, year, habitat, trophic level, phylum or other broad classification of organism studied, variables measured and other relevant details, if necessary. If the study was done in other than natural conditions, these are indicated. Experimental results are given under two experimental types: type 1, occurrence of interspecific competition, and type 2, relative strength of intraspecific competition vs. interspecific. Type 1: For each response studied (e.g., growth, mortality, etc.), the fraction listed for each species (or group of species as indicated) is the number of experiments in which interspecific competition was found/the total number of experiments performed. If data on temporal or spatial variation exist, the fraction indicates the number of experiments in different years (yr), seasons (t), or sites (s). If in a pair of species density was manipulated for both, the results for the pair are given as follows: If one responded and the other did not, this asymmetrical result is indicated by "asym."; if both responded, it is so indicated; if neither responded, it is so indicated. These responses refer only to ones indicating competition; a response indicating a positive interaction was scored as no competitive response, and these are indicated separately. Type 2: For each taxon studied, the strength of intraspecific competition relative to interspecific with every other taxon is given for each variable studied. The meanings of the abbreviations used are, e.g., "Sp. X, intra > with Y" indicates that for sp. X, intraspecific competition is stronger than interspecific competition with sp. Y; "inter > from Y" indicates that interspecific competition from sp. Y is stronger than intraspecific; "intra = with Y" indicates that the two types of competition are equivalent. If neither type is detectable, no entry is given. If there are only two species, the species abbreviations are omitted as being unnecessary. Exp. = experiment(s).

Some papers with field experiments on competition in the journals and years examined were not included for the following reasons: (1) The setup of the experiment bore little apparent resemblance to any existing natural situation, and no data were provided to indicate whether such a resemblance exists. (Such arrangements are properly designed to test certain hypotheses, but it is very difficult to know whether the results can be extended to nature.) (2) In some cases it was impossible to decide whether the effect could be caused by interspecific competition. A group of species was manipulated together and the subsequent response of the same group observed. Thus the responses seen may have resulted from either intraspecific or interspecific interactions. (3) Instead of providing the analyzed or summarized data, only the probability level of the significance test was given so that it is not possible for the reader to evaluate the results. (4) If the species potentially could compete, but the interaction was a predator-prey one, some workers regard this as an extreme form of interference. However, I did not include any papers in this category. (5) If controls were absent, or were not done at the same time as the treatments, or if the control sites had different environmental conditions from the treatments, these were excluded.

1. Abramsky et al. 1979. Terrestrial mammal omnivore density on irrigated and fertilized field. Type 1: 1 sp. 1/1.
2. Abramsky and Sellah 1982. Terrestrial mammal granivore density. Type 1: 1 sp. 0/1. No positive response.
3. Adams and Traniello 1981. Terrestrial insect (omnivore), effect on immigration of one sp. to food (bait) guarded by another sp. Type 1: 1/1, short-term behavioral response.
4. Allen and Forman 1976. Terrestrial plants, herbs on three 6-yr old fields, density. Type 1: all exp.: *Potentilla* 3/4, *Convolvulus* 2/11, *Aster* 0/11, *Hieracium* 0/8; 3 other spp., each 0/4; 2 other spp., each 0/3. Of these, the following had data on spatial variation: 2 pairs of spp. on each of 2 different sites, 4 @ 0/2 s; 1 pair of spp. on 3 sites, 2 @ 0/3 s. Pairs: 17 neither, 5 asym.: *Aster* > *Potentilla*, *Fragaria* > *Convolvulus*, *Solidago* > *Convolvulus*, *Hieracium* > *Potentilla*, *Plantago* > *Potentilla*; no reversals within pairs. Positive response: *Aster* 1/11.
5. Bertness 1981. Marine omnivorous crustacea, crabs transplanted to pools, short-term behavior (emigration), 2 spp. Type 1: 1/1, 0/1. Pair: asym. No positive response.
6. Black 1979. Marine molluscan herbivore density, intertidal shore height, 1 sp. Type 1: 1/1.
7. Brown 1982. Freshwater herbivorous snail fecundity and growth, 2 spp. in 1-liter containers in 3 ponds. Type 1: *Physa* fecundity 3/3 s, growth 0/3 s; *Lymnaea* fecundity 3/3 s, growth 1/3 s. No positive responses. Pairs: fecundity both; growth asym., no reversals. Type 2: *Physa* fecundity inter >, 3 ponds; growth intra >, 3 ponds; *Lymnaea* fecundity intra >, 1 pond, intra =, 2 ponds; growth intra >, 1 pond, intra =, 2 ponds.
8. Chappell 1978. Terrestrial omnivorous mammals, shift in altitudinal distribution on mountain, 2 spp. Type 1: 0/1, 1/1. Pair: asym. No positive response.
9. Creese and Underwood 1982. Marine herbivorous snails in cages, mortality (mort.), growth, and weight (wt.). $C.$ = *Cellana*, $S. v.$ = *Siphonaria virgulata*, $S. d.$ = *S. denticulata*. For mort., exp. were done in the following 2 periods: May–Aug. 1977 with all 3 spp., July 1979–Jan. 1980 with $C.$ and $S. d.$ only. For growth, exp. were done in either 2 or 3 separate areas for the above 2 periods, respectively, giving a total of 5 exp. for $C.$ and $S. d.$ and 2 exp. for $S. v.$ For weight, 3 exp. in the second period for $C.$ and $S. d.$ For mixed spp. densities of 20, treatments with 15 $C.$ were not included since treatments could not be maintained because of high mort. Type 1: After each fraction is shown the other sp. in the pair. $C.$: mort. 0/1$S. v.$, 0/2$S. d.$; growth 0/2$S. v.$, 0/5$S. d.$; wt. 0/3$S. d.$ $S. v.$: mort. 1/1$C.$, 0/1$S. d.$; growth 2/2$C.$, 2/2$S. d.$ $S. d.$: mort. 2/2$C.$, 0/1$S. v.$; growth 5/5$C.$, 2/2$S. v.$; wt. 3/3$C.$ No positive responses. Temporal (seasonal) variation (same for both mort. and growth): 2 @ 0/2 t, 2 @ 2/2 t; spatial variation (growth only): 2 @ 0/2 s, 4 @ 2/2 s, 0/3 s, 3/3 s. Type 2: All comparisons are between single sp. vs. 2 spp. treatments at the same total density. $C.$: mort., intra > with $S. v.$, 1 exp., intra > with $S. d.$, 2 exp.; growth, intra = with $S. v.$, 2 exp., intra = with $S. d.$, 5 exp.; wt. intra > with $S. d.$, 3 exp. $S. v.$: mort., inter > from $C.$, 1 exp., intra = with $S. d.$, 1 exp.; growth intra = with $C.$, 2 exp., intra = $S. d.$, 2 exp.; $S. d.$: mort., inter > from $C.$, 2 exp., intra = with $S. v.$, 1 exp.; growth, intra = with $C.$, 5 exp., intra = with $S. v.$, 2 exp.; wt. intra = with $C.$, 3 exp. Pairs: mort.: $C. > S. v.$ asym., $C. > S. d.$ asym.; $S. v./S. d.$ neither; growth $C. > S. v.$ asym., $C. > S. d.$ asym. $S. v./S. d.$ both; wt. $C. > S. d.$ asym. No reversals within pairs.
10. Dayton 1975. Marine algal cover (figs. 3, 4 only, since in other exp. [figs. 1, 2] algal removal was confounded with herbivore departure). Type 1: Spatial variation, 2 sites: fugitive spp. when *Hedophyllum* (*H.*) removed 1/2 s; with *Lessoniopsis* (*Le.*) removed 2/2 s; *H.* when *Le.* removed 2/2 s. With no data on variation: fugitive spp. when *Laminaria* (*La.*) removed 0/1, *H.* with *La.* removed 0/1, *La.* when *H.* removed 0/1, *La.* when *Le.* removed 1/1; other ephemeral spp. when *Le.* removed 1/1; obligate understory spp. when *H.* removed 0/1. Total by sp.: fugitive 1/2 s, 2/2 s, 0/1; *H.* 2/2 s, 0/1; *La.* 0/1, 1/1; other ephemeral spp. 1/1, obligate understory 0/1. Pairs: (fig. 4) *Le.* > *H.* and *Le.* > *La.* both asym.; *H./La.*, neither. Positive response: obligate understory to *H.* 1/1.
11. DeBenedictis 1974. Freshwater omnivorous larval amphibians of 2 spp. in pens; 4 responses measured were: survival, duration of larval period, body length, and biomass at metamorphosis. Type 1: Competition found when predators excluded but no competitive effects in open pens with predators; 2 yr, 2 spp. @ 0/2 yr for each of the 4 responses. Pairs: neither for all responses. No positive responses.
12. Dhondt and Eyckerman 1980. Terrestrial omnivorous bird density: 1 sp. Type 1: no variation in 2 yr or between 2 sites in 1 yr: 3/3.

13. Duggins 1980. Marine algal cover, taxon is a group of annuals. Type 1: 1/1.
14. Duggins 1981. Marine herbivorous echinoderms, effect of removal or addition of 1 sp. on density of 2 other spp. Type 1: density 0/2, 0/2; gonad index 0/2, 0/2. No pairs. Positive responses: density 1/2; gonad index 1/2, 0/2.
15. Dunham 1980. Terrestrial carnivorous lizards, *Urosaurus* (*U.*) and *Sceloporus* (*S.*), for up to 4 yr. Type 1: Data on annual variation: density: *U.* 2/3, *S.* 0/3; first yr survival *U.* 1/3, *S.* 1/3; adult survival *U.* 0/3, *S.* 2/6; growth *U.* 2/4, *S.* 0/4. Seasonal variation: foraging rate *U.* 1/2, *S.* 0/2; no data on variation: body lipids *U.* 2/2, *S.* 0/2; body mass *U.* 2/2, *S.* 0/2. Pairs: 5 variables asym. *S.* > *U.*, 1 variable asym. *U.* > *S.*, first yr survival both. Positive responses: first yr survival *U.* 1/3, adult survival *U.*2/3.
16. Eriksson 1979. Aquatic predaceous bird density, competition with fish. Type 1: annual variation, (bird) 2/3 yr. No positive responses.
17. Fonteyn and Mahall 1981. Terrestrial plants in desert, effect of removal of each sp. on xylem pressure potential of other sp. Type 1: Seasonal variation, *Larrea* 2/2, *Ambrosia* 2/2. Pairs: 2 seasons, both. Type 2: *Larrea* autumn 1977 inter >, spring 1978, intra = ; *Ambrosia* autumn 1977 and spring 1978, both inter >. Note: table 3 in Fonteyn et al. (1981) contains typographical errors; the above correct results were given me by the authors.
18. Fowler 1981. Terrestrial plant percentage cover on field regularly mown for 30 yr; 8 spp. removed one at a time; no data used for removals of groups of spp. Type 1: *Plantago* (*Pl.*) 1/6, *Poa* 2/7, *Rumex* 0/7, *Anthoxanthum* 1/3, *Trifolium* 0/3, *Allium* 0/3, *Cynodon* (*Cyn.*) 1/7, *Salvia* 1/8, *Carex* 0/4, *Setaria* 1/5, *Paspalum* sp. 1/3, *Paspalum dilatatum* (*Pa. d.*) 2/4, *Paspalum laeve* (*Pa. l.*) 0/4, *Paspalum ciliatifolium* 0/5, winter annuals 1/3. Total 72 exp. Pairs: April: 3 pairs: *Rumex* > *Pl.* asym.; *Pl./Poa* neither, *Rumex/Poa* neither. Sept.: 3 pairs asym.: *Pl.* > *Pa. d.*, *Pl.* > *Cyn.*, *Cyn.* > *Pa. d.* (a transitive hierarchy of these 3 spp.). 7 pairs neither: *Pl./Pa. l.*, *Pl./Carex*, *Pa. d./Pa. l.*, *Pa. d./Carex*, *Pa. l./Cyn.*, *Pa. l./Carex*, *Cyn./Carex*. Total asym. 4; neither 9. Positive responses, 1 each: *Allium* to *Poa*, *Carex* to *Pa. l.*, *Rumex* to *Pa. d.*
19. Friedman and Orshan 1974. Terrestrial plants in desert, effect of removal of other spp. on fecundity of 2 varieties of a different sp., 2 sites. Type 1: spatial variation: no irrigation 0/2 s, 0/2 s; with experimental irrigation, 1/2 s, 1/2 s. Pairs: neither on both sites. No positive responses.
20. Friedman et al. 1977. Terrestrial plants in desert, taxon is one group of annual spp. on 2 sites, response to removal of bushes of 2 spp. Type 1: density 2/2 s, weight 1/2 s. No pairing. No positive responses.
21. Grace and Wetzel 1981. Freshwater plant (cattails) productivity, 2 spp. Type 1: 1/1, 0/1. Pair: asym. No positive response.
22. Gross 1980. Terrestrial plants on third-yr old field, effect of removal of 3 spp. groups, singly or in combination, on the density and survival of another species, 3 cohorts of which emerged in 3 different seasons. Diffuse competition effects as follows. Type 1: survival to reproduction, 3 separate cohorts 2/3 t; seedling emergence, cohorts combined 1/1; juvenile survival 3 separate cohorts 3/3 t. No pairing. No positive responses.
23. Gross and Werner 1982. Terrestrial plants, 2 sites, herbs on old field; effect of presence of existing vegetation on establishment and survival of 4 spp. Although *Daucus* and *Tragopogon* are among the dominant sp. in 15-yr old field, they were rare in the area of the study plots, so that competition was mainly interspecific (K. Gross, personal communication). Type 1: seedling emergence: *Verbascum* 2/2 s, *Daucus* 2/2 s, *Tragopogon* 2/2 s. Seedling emergence: *Oenothera* 1/1 (table 4). Rosette survival: *Verbascum* 1/2, *Oenothera* 1/2, *Daucus* 1/2 s, *Tragopogon* 1/2 s (tables 5, 8). No pairing. No positive responses.
24. Hairston 1980*b*. Terrestrial salamanders (predators), 2 spp., 2 sites. Type 1: density, *Plethodon glutinosus* 2/2 s, *P. jordani* 0/2 s; age structure, *jordani* 2/2 s. Spatial variations in the rate of response, faster where altitudinal overlap of 2 spp. is least. Pair: density asym. No positive response.
25. Hairston 1981. Terrestrial salamander (predators) density. Type 1: response of 4 spp. to removal of 2 other spp. separately, all at 2 sites: 4 spp. @ 0/2 s, 0/2 s. No pairs. No positive responses.
26. Hils and Vankat 1982. Terrestrial plant biomass, herbs on first-yr old field; effects of 6 types of spp. removals on 3 taxa (annuals, biennials, perennials). Type 1: 0/2, 0/5, 0/4, respectively. No pairing. No positive responses.
27. Hixon 1980. Marine fish (predators), shift in depth distribution; exp. when fish alone removed.

Type 1: No data on variation: *Embiotoca lateralis* 0/1; data on annual variation *E. jacksoni* 2/2 y. Pair: asym. No positive response.
28. Holbrook 1979. Terrestrial omnivorous mammal activity level and resource use. *P. b.* = *Peromyscus boylii*, *N.* = *Neotoma*, *P. m.* = *Peromyscus maniculatus*. Type 1: Activity level: *P. b.* 0/1, *N.* 0/1, *P. m.* 2/2; vegetation use; *P. b.* 0/1, *N.* 1/1, *P. m.* 1/2; arboreal activity: *P. b.* 0/1, *N.* 1/1, *P. m.* 2/2. Pairing of *P. b.* and *N.*: activity neither, vegetation use asym., arboreal use asym. Positive response, vegetation use *P. b.* 1/1.
29. Inouye, D. 1978. Terrestrial insects (nectivores), 2 spp., 2 yr for 1 sp. Type 1: density, short-term interference 1/1, 2/2 yr. 1 pair, both.
30. Inouye, R. 1981. Terrestrial herbivorous fungal parasite of plants (attacks leaves, flowers, fruits); effect of removal of granivorous rodent competitor, 1 sp. Type 1: 1/1.
31. Kastendiek 1982. Marine coelenterate (predator). Type 1: nos. 1 sp. 1/1.
32. Kroh and Stephenson 1980. Terrestrial plant weight, on first-yr fallow old field. Estimated only the relative strength of intraspecific competition vs. interspecific. *A.* = *Amaranthus*, *C.* = *Chenopodium*, *P.* = *Panicum*, *S.* = *Setaria*. Results: *A.*: intra = with *C.*, intra > with *P.* and *S.*; *C.*: intra = with *A.*, intra > with *P.* and *S.*; *P.*: intra = with *S.*, inter > from *A.* and *C.*; *S.*: intra = with *P.*, inter > from *A.* and *C.*
33. Larson 1980. Marine fish (predators), shifts in depth distribution. *Type 1:* 2 spp., 1/1, 1/1. Pair: both.
34. Lubchenco 1980. Marine algal distribution, effect of removal of 1 sp. on 2 other spp.; % cover, 2 sites. Type 1: 2/2 s, 2/2 s. No pairing.
35. Lynch 1978. Freshwater suspension-feeding herbivorous crustaceans in containers in pond; summer exp., nos. of *Daphnia* (*D.*) and *Ceriodaphnia* (*C.*), autumn exp. nos. and clutch size of *D.*, *C.*, and *Bosmina* (*B.*). Type 1: nos.: *D.* with *C.* 1/2 t, *D.* with *B.* 0/1; *C.* with *D.* 1/2 t, *C.* with *B.* 1/1; *B.* with *D.* 1/1, *B.* with *C.* 1/1; clutch size all 3 spp. 0/2. Pairs: *D./C.*: summer *C.* > *D.* asym., autumn *D.* > *C.* asym., *D.* > *B.* asym., *C./B.* both. Total 2 asym., 1 both. Positive responses: exp. 2 autumn, *D.*: nos. 2/2, clutch 2/2. Intransitive network of the 3 spp.
36. Mackie et al. 1978. Freshwater suspension-feeding herbivorous molluscs in containers in pond; reproductive rate. *M. t.* = *Musculium transversum*, *M. s.* = *M. securis*. Type 1: 1/1, 1/1. Pair: both. Type 2: *M. t.*: inter > 1 exp.; *M. s.* inter > 1 exp.
37. McClure and Price 1975. Terrestrial (herbivorous) insects, no. of young produced per female at different densities of adults (equal nos. of both sexes) in cages on tree leaves; abbreviations as in paper. Type 1: *Erythroneura ingrata* (I) and *E. bella* (B) 1/1; *E. morgani* (M) and *E. torella* (T): 2/2; *E. arta* (A) and *E. usitata* (U): 3/3; *E. lawsoni* (L): 4/4. 8 pairs, all both. Positive responses at lowest densities: B to A, L to T, T to L, M to T, T to M, M to U. Type 2: A: intra = with L, intra > with U, intra > with B; U: intra = with L, intra > with A, intra > with M; B intra > with A; L intra = with A, I and U, intra > with T; M intra > with T and U; I intra = with L; T: intra > with M, intra = with L.
38. McLay 1974. Freshwater plant density, 1 sp. Type 1: 1/1.
39. Menge 1976. Marine invertebrate (suspension-feeding omnivore) abundance, 1 sp. in cages at 13 sites where both competitors alone can reach 100% cover. Type 1: 4/13 s. No pairing. No positive responses.
40. Miles 1974. Terrestrial plant seedling establishment, 8 spp.; with soil bared and unfertilized, results on 4 sites for each sp.; one site modified by grazing. Type 1: *Agrostis* 2/4 s, *Deschampsia* 4/4 s, *Holcus* 4/4 s, *Hypochoeris* 2/4 s, *Luzula* 3/4 s, *Rumex* 2/4 s, *Sarothamnus* 1/4 s, *Ulex* 0/4 s. No pairing.
41. Minot 1981. Terrestrial bird (omnivore) density, 1 sp. Type 1: 1/1.
42. Montgomery 1981. Terrestrial omnivorous mammals, 2 spp.; first fraction *Apodemus sylvaticus*, second *A. flavicollis*. Type 1: density 0/1, 1/1; survival 0/1, 0/1; immigration 0/1, 1/1; breeding season 1/1, 1/1; body weight (sexes separate) 1/2, 1/2; microhabitat use 1/1, 1/1. Pairs: 2 responses asym., 1 neither, 4 responses both.
43. Morse 1981. Terrestrial nectivorous insect numbers. Type 1: short-term interference 1/1.
44. Peterson 1979. Marine crustacean (suspension-feeding omnivore) abundance, 1 sp. Type 1: 1/1.
45. Peterson and Andre 1980. Marine suspension-feeding herbivorous mollusc growth in enclosures, 2 yr, results indicate which competitor was removed. *S.* = *Sanguinolaria*, *P.* = *Protothaca*, T = 2 other spp. Type 1: *S.* 2/2 yr (T), 0/2 yr (*P.*); *P.* 0/2 yr (T), 0/2 yr (*S.*). Pair: *S./P.* neither both yr.

46. Peterson 1982. Marine suspension-feeding herbivorous molluscs in enclosures; 2 spp., 3 sites, 2 yr, 2 seasons. *P.* = *Protothaca*, *C.* = *Chione*. Type 1: Growth: *P.* 1/12, *C.* 0/11; recruitment *P.* (3 seasons, 3 sites) 0/9; *C.* (1 season and site) 1/1; proportion of body mass as gonad (1 site, 1 season) *P.* 1/1, *C.* 0/1. Pairs: for 2 variables asym. *C.* > *P.*; 1 variable asym. *P.* > *C.* Positive responses: growth, *P.* 1/12, *C.* 1/11. Type 2: Small *P.* growth intra > 6 exp.; *C.* growth intra > 2 exp. No competition in other 15 exp. *P.* recruitment intra > 1 exp. no competition 8 exp. *C.* recruit inter > 1 exp. Gonad proportion, *P.* intra =, *C.* intra >.
47. Petranka and McPherson 1979. Terrestrial plant, seedling germination in forest-prairie ecotone. Taxon is group of winter annuals; effects of adding shading, toxins, and litter. Type 1: seasonal variation shading exp. 3/3 t; no data on variation; toxins exp. 1/1. No pairing.
48. Pinder 1975. Terrestrial plant productivity in 18th-yr old field; taxon a group of forb spp. Type 1: 1/1.
49. Price 1978. Terrestrial granivorous mammals in pens, microhabitat use; results included here only where there was a contemporaneous control. 3 spp. of *Perognathus* combined, table 4. Type 1: 1/1.
50. Race 1982. Marine mollusc (deposit-feeding omnivore) distribution, 1 sp. Type 1: numbers, 1/1.
51. Rahman 1976. Terrestrial plant growth, 2 spp., 3 sites; estimated only the relative strength of intraspecific competition vs. interspecific; 1969 exp. in planted plots in grassland. Results: *Dactylis* intra >, 1 exp.; intra =, 2 exp.; *Deschampsia* inter >, 2 exp.; intra =, 1 exp.
52. Raynal and Bazzaz 1975. Terrestrial plant seed production in first-yr old field, height, weight, 1 sp., 3 yr. Type 1: 3/3 yr for each response; no temporal variation. No pairing.
53. Redfield et al. 1977. Terrestrial omnivorous mammal density and date of onset of breeding season, 1 sp. Type 1: 1/1 for each response.
54. Robertson et al. 1976. Marine herbivorous fish, feeding activity. Type 1: interspecific competition: effect of removal of pomacentrid on: parrotfish 1/1; nonparrotfish 4 spp. @ 1/1. Type 2: parrotfish intra > inter with pomacentrid. No pairing.
55. Schroeder and Rosenzweig 1975. Terrestrial granivorous mammal habitat use, 2 spp. Type 1: numbers, 0/1, 0/1. Pair: neither.
56. Seifert and Seifert 1976. Terrestrial herbivore and detritivore insect density; 4 spp. insects matched in pairs in each of 2 spp. *Heliconia* plants; abbreviations use first initials of each insect genus name. *Cephaloleia* (*Ce.*), *Gillisius* (*G.*), *Beebeomyia* (*B.*) are herbivores; *Quichuana* (*Q.*), *Copestylum* (*Co.*), *Merosargus* (*M.*) feed on detritus and nectar. (Although some of these insects live underwater in the flower bracts, they are classed as terrestrial because they live on a land plant.) Type 1: *Heliconia wagneriana* (*H. w.*): *Q.* 1/3, *G.* 0/3, *Co.* 0/3, *B.* 1/3; *Heliconia imbricata* (*H. i.*): *Q.* 0/3, *G.* 2/3, *Ce.* 1/3, *M.* 0/3. Same insect sp. (*Q.* and *G.*) in both plants: in *H. w.* no competition, in *H. i. Q.* responded positively; *G.* responded negatively. Positive responses: *Q.* to *B.*, *Q.* to *G.*, *Ce.* to *M.*, *Ce.* to *Q.*, *M.* to *Ce.* Type 2: *H. w.*: *Q.* inter > from *Co.*; *G.* intra > with *Q.*, *B.*, *Co.*; *Co.* intra > with *Q.*, *G.*, *B.*; *B* intra > with *Q.*, *G.*, *Ce. H. i.*: *G.* inter > from *Q.*, *M.*; *Ce.* inter > from *G.*; *M.* intra > with *Q.*, *G.*, *Ce.* Same insect pair in different plants: *H. w.*: *G.* intra > with *Q.*; *H. i.*: *G.* inter > from *Q.* No competition of either type, 8 other exp. Pairs: 5 asym.: *Co.* > *Q.*, *G.* > *B.*, *Q.* > *G.*, *M.* > *G.*, *G.* > *Ce.*; 6 neither: *Q./B.*, *G./Co.*, *Co./B.*, *Q./M.*, *Q./Ce.*, *M./Ce.*
57. Seifert and Seifert 1979. Terrestrial herbivore and detritivore insect density; 4 spp. insects matched in pairs in one sp. *Heliconia* plant. *Cephaloleia* (*Ce.*) and *Gillisius* (*G.*) are herbivores; *Quichuana* (*Q.*) and *Copestylum* (*Co.*) eat detritus and nectar. (Although some of these insects live underwater in the flower bracts they are classed as terrestrial because they live on a land plant.) Type 1: *Q.* 0/3, *G.* 0/3, *Co.* 0/3, *Ce.* 1/3. Positive response, *G.* to *Ce.* Type 2: *Ce.* intra > with *Q.*, *G.*, *Co.*, no competition, 9 other exp. Pairs: 1 asym., *Co./Ce.* 5 neither: *Q./G.*, *Q./Co.*, *Q./Ce.*, *G./Co.*, *G./Ce.*
58. Smith 1981. Terrestrial carnivorous lizards; each fraction is the no. yr for each sp. (age or sex sometimes separated): *S.* = *Sceloporus*, *U.* = *Urosaurus*. Type 1: Nos. all ages *S.* 0/3, *U.* 0/3; survival young females *S.* 1/2, *U.* 1/2; young males *S.* 0/2, *U.* 0/2; older males and females *S.* 0/2, 0/2; *U.* 0/2, 0/2; growth (only 3 groups that grew) *S.* young females 1/2, young males 0/2, *U.* young females 0/2. Pairs: neither for all variables except asym. for growth of young females and both for survival of young females; no reversals within pairs.
59. Smith and Cooper 1982. Freshwater suspension-feeding herbivorous crustaceans in containers in a

pond, population growth. Type 1: exp. 2, July (3 spp): *Daphnia* (*D.*) 2/2, *Moina* (*M.*) 1/2, *Ceriodaphnia* (*C.*) 2/2; exp. 3, Aug. (2 spp.): *D.* 1/8, *C.* 7/8. No positive responses. Pairs: exp. 2 *D./M.* both, *D./C.* both, *M.* > *C.* asym. (Intransitive network of 3 spp., no reversals within pairs). Exp. 3 *D./C.* asym. Temporal variation *D.* 1/2 t, *C.* 2/2 t. Type 2: exp. 3, D., intra >, *C.* inter >.

60. Sousa 1979. Marine algal cover. Type 1: *Ulva* removal: *Gigartina* (*Gi.*) sp. 1/1; *Gi. leptorhynchos* (*Gi. l.*) removal: *Ulva* 1/1, *Gi. canaliculata* (*Gi. c.*) 1/1, *Laurencia* (*L.*) 1/1, *Gelidium* (*Ge.*) 0/1; *Ge.* removal: *Ulva* 1/1, *Gi. c.* 1/1, *Gi. l.* 0/1, *L.* 0/1. *Gi. c.* removal: *Ulva* 1/1. Totals: *Ulva* 3/3, *Gi. c.* 2/2, *Gi. l.* 0/1, *Gi.* sp. 1/1, *L.* 1/2, *Ge.* 0/1. Pairs: *Gi. l./Ge.*, neither. No positive responses.

61. Strong 1982. Terrestrial herbivorous insect density, effect on 3 invading sp. of varying the densities of either *consanguinea* or *perplexa* in *Heliconia* host plant. Type 1: invaders: *vicina* 0/2, *consanguinea* 0/1, *perplexa* 0/1. Pairs: *consanguinea/perplexa* neither. Positive response of *perplexa* to *consanguinea*.

62. Sutherland 1978. Marine sessile omnivorous suspension-feeding invertebrates on undersides of fouling panels hung below a dock; removed *Schizoporella* (*Sc.*) or *Styela* (*St.*) one at a time in each of 2 yr. Type 1: Data on annual variation in % cover: 5 spp. (*Balanus, Halichondria, Haliclona, Sc., St.*) @ 1/2 yr; 4 spp. (*Bugula neritina, Ascidea, Botryllus, Tubularia*) @ 0/2 yr. No data on annual variation, 7 spp. @ 0/1. Pairs: *Sc.* and *St.* asym. 1 yr. Positive responses: *St.* to *Sc.* 1/2 yr, *Hydroides* to *Sc.* 1/2 yr.

63. Taylor and Littler 1982. Marine algae, herbivorous grazers and sessile suspension-feeding omnivorous animal abundance; effect of removal of sea anemone on % cover of algae and *Phragmatopoma*, and on no. of barnacles. Type 1: algae 3 spp. @ 1/1, 2 spp. @ 0/1; animals 2 spp. @ 1/1. No pairing. Positive responses of 2 spp. algae to anemone.

64. Tinkle 1982. Terrestrial predatory lizard, 1 sp. 2 yr. Type 1: With data on annual variation: density 0/2 yr, survival 0/2 yr, yearling body size 0/2 yr, older body size 0/2 yr. Without data on variation: habitat use 0/1. No pairs. No positive responses.

65. Turkington et al. 1979. Terrestrial plant, 1 sp. on 4 sites each having a different species of competitor, in pasture grazed by sheep for 60 yr. Type 1: seed sown, seedling emergence 4/4; transplanted seedlings, survival 4/4, yield 4/4. No pairing.

66. Underwood 1978. Marine herbivorous snails in cages, mortality (mort.) and weight (wt.) change. *C.* = *Cellana*, *N.* = *Nerita*, *B.* = *Bembicium*. Experiments: 100 days all 3 spp., 200 days, *N.* and *B.* only; each fraction refers to exp. between a particular pair of species shown by the abbreviations. The number of experiments for each species is as follows: for mortality, 1 exp. replicated in time; for wt., 2 exp. at different seasons, each having as replicates the different individuals measured. The author provided the analyses for the 2 different times, since these data had been pooled in the published paper. Type 1: *C.*, mort. 1/1 *N.*, 0/1 *B.*, wt. 2 seasons: 2/2 *N.*, 0/2 *B.*; *N.*: mort. 0/1 *C.*, 0/2 *B.*, wt. 0/2 *C.*, 0/2 *B.*; *B.*: mort. 1/1 *C.*, 1/2 *N.*, wt. 2/2 *C.*, 2/2 *N.* No positive responses. Temporal (seasonal) variation in wt.: *C.*: 0/2, 0/2; *N.*: 0/2, 0/2; *B.*: 2/2, 2/2. Type 2: *C.*: mort., intra = with *N.*, intra > with *B.*; wt., intra = with *N.*, intra > with *B.*; *N.*: mort., 200 days intra > with *B.*, wt., 200 days intra > with *B.*; *B.*: mort., 100 days inter > from *C.*, 200 days intra = with *N.*; wt. 100 days inter > from *C.*, 200 days intra = with *N.* Pairs: mort.: *N.* > *C.* asym., *N.* > *B.* asym. or neither, *C.* > *B.* asym.; growth: *N.* > *C.* asym., *N.* > *B.* asym., *C.* > *B.* asym. Three-species transitive hierarchy; no reversals within pairs.

67. Werner, E., and Hall 1977. Freshwater predaceous fish, juveniles only, placed in experimental ponds, 2 spp., estimated only the relative strength of intraspecific competition vs. interspecific. Results: survival, intra = for both spp.; growth, bluegill inter >, green sunfish intra = ; food type, intra = for both spp.; food size, intra = for both spp.

68. Werner, P. 1977. Terrestrial plant productivity in second-yr and third-yr fallow old fields. Effect of experimental introduction of a different colonizing sp. on 2 taxa (grasses, herbaceous dicots); 3 sites, 2 yr. Type 1: var. space: grasses 0/3 s; variation time: grasses 0/2 yr, dicots 2/2 yr. Total: grasses 0/6, dicots 2/2. No pairing. No positive results.

69. Williams 1981. Marine herbivorous echinoderm density, 2 spp. Type 1: removed fish, 1/1, 1/1; removed other sp. of echinoid 1/1, 1/1. Total 2/2, 2/2. One pair, both.

70. Wise 1981a. Terrestrial predatory spiders in open frames, density, 2 spp. *B.* = basilica (*Mecynogea lemniscata*), *L.* = labyrinth (*Metepeira labyrinthea*). Type 1: all 4 variables (see below) 0/1 for each sp. Type 2: survival, *L.* intra >; rate of egg production, no competition either sp.; eggs per sac, *B.* intra >. Web height *B.* intra >. Pair: neither. No positive responses.

71. Wise 1981b. Terrestrial omnivorous insect density, 3 yr, 4 spp., effect of removal of fifth sp. Type 1: 4 spp. @ 0/3 yr. No pairs. Some positive responses which were inconsistent throughout a yr.
72. Woodin 1974. Marine deposit-feeding omnivorous annelid density in cages. Type 1: Burrowing sp. increased when the total of 3 spp. of tube-builders was reduced at 2 depths: 2/2 s. No pairing.

APPENDIX B

On Testing Competition Theory

The character and intensity of biological interactions have undoubtedly been modified during evolution. Direct evidence of such coevolution comes from studies of parasite-host interactions between pathogens and their plant hosts (Van der Plank 1968). Similar coevolution may have occurred between species of competitors, reducing the intensity and frequency of interspecies competition.

In a recent paper (Connell 1980) I discussed the adequacy of the existing evidence concerning the coevolution of competitors. Contrary to the statements in Pacala and Roughgarden (1982, p. 444) and Roughgarden (1983, p. 583), my paper did not criticize competition theory; it was critical only of the adequacy of the evidence that had been marshalled to test the theory of coevolution of competitors. As did Grant (1972, 1975), I concluded that the existing evidence was weak. I also argued that coevolution would be more likely to occur between populations on different trophic levels (predator-prey, parasite-host, etc.) than between competitors. My reasoning was that since predators or parasites are dependent upon their hosts for food, there would be continual interactions as the natural enemies sought out and fed on their prey. In contrast, since neither competitor is dependent upon the other, there would be less reason for the two to co-occur, so that one would expect more variability in this interaction than between natural enemies and their prey. Since variability in the interaction has been directly observed to reduce the possibility of coevolution between wheat and the pathogenic rust attacking it (Van der Plank 1968), we might also expect the same reduction in selection pressure for coevolution between competitors when competition is variable. The evidence given in the present paper suggests that interspecific competition is very variable.

Since direct evidence is preferable to reasoning by analogy, I suggested (Connell 1980) some designs for field experiments to test for competitive coevolution. My reasoning and experimental suggestions have been criticized by Roughgarden (1983) on several grounds. First, he argues that no single experimental protocol can guarantee that competition actually occurs in a system because the circumstances, mechanisms of competition, and population structure all vary. Since the success of any experiment depends upon how well it is executed, as discussed above, I agree that there is no guarantee that an experiment that followed mine or any other single protocol will demonstrate the existence of competition. Also, since the intensity of competition varies, a single experiment may fall in a period when competition intensity is very weak, and so give negative results. However, the point of doing replicated experimental manipulations with associated controls is to reduce the effects of the external contingencies and environmental variations by including them in the experiment. Properly arranged treatments and controls are open to the same degree of environmental variation, and replication is done to ensure that the results are not heavily influenced by a single extreme case, as described earlier. So within an experiment, variation is controlled for, given sufficient replication.

Roughgarden (1983) makes four other detailed criticisms of my experimental protocol; I will refer to them by the letter designations used by him. Point A: Divergence through time is only one possible outcome of coevolution, and a rather unlikely one. Possibly so, in theory; however, the theoretical argument given by Roughgarden to support this point specifies no assumptions or equations, so that there is no way to judge its applicability to reality. In any case, I stated that my protocol applies only to populations that show such divergence; they are the ones most often used to illustrate competitive coevolution and seemed to me to provide the best situation to use for a field experimental test. (The same uncertainties apply to his other use [pp. 589, 590] of a theoretical argument against my contention that predators are effective in some conditions in reducing competition by keeping populations below the level at which they compete. In any case, his argument does not apply to the situation described in my paper, since in it, prey populations do compete.) Point B: The protocol focused on niche compression and expansion, not on niche shifts. This is true; elsewhere in my paper I had referred to niche divergence; it is easily included by adding the appropriate wording. Point C suggests that, since competition is greatly reduced after coevolutionary divergency has occurred, it would tax the resolution of my experimental protocol so the recommended test is unworkable. However, this is wrong; my experiments were designed precisely to address this problem, by transplanting the allopatric population that had not yet diverged. Point D states that detecting a genetic basis is difficult. I agree, but nothing in the subsequent discussion shows that the method I suggested would not be effective. Roughgarden's remarks (pp. 589, 590) on the problems of interpreting experimental results from intertidal systems are well taken, but I was not proposing that such intertidal systems be used to test the theory. Rather the example (of J. R. E. Harger's work) was used because it was the only one I knew in which both a transplant from allopatry to sympatry and a field experiment on competition had been done.

Roughgarden (p. 590) states that my experiments are "biased toward the detection of interference mechanisms." While this may be so, it should not actually introduce bias, since he also suggests that "exploitative competition should cause the evolution of interference mechanisms." If so, then detecting interference mechanisms should also provide indirect evidence of exploitation. He states that the lack of direct evidence concerning exploitative mechanisms "is an artifact of studies focused on the observation of interindividual interactions." Whether this lack indicates that exploitation is actually of little importance or is an artifact of improper focus, no one knows.

To sum up, I feel that whether the theory concerning the coevolution of competitors is true or not, it has only begun to be adequately tested in the field. Roughgarden (1983, pp. 593, 583) states that his paper is a response to the "antitheoretical rhetoric" in Connell (1980) and two other papers. "These criticisms imply that competition theory, including its extension to the coevolution of competitors, is irrelevant to natural processes and is unworthy of testing regardless of whether the testing is feasible." In regard to my paper, these assertions are wrong; I did not criticize competition theory nor imply that it is unworthy of testing; in fact I proposed experimental tests. The only way I know to decide whether competition theory is relevant or irrelevant is to formulate testable hypotheses and test them, using either field experiments or statistical analyses such as Dunham et al. (1979) have done.

Ecological theory does not establish or show anything about nature. It simply lays out the consequences of certain assumptions. Only a study of nature itself can tell us whether these assumptions and consequences are true. My attitude to ecological theory is not to suggest that theory as such is worthless, but to test it.

Roughgarden also states that mine and the other two papers "criticize the evidence for competition" (p. 583) and that my paper "maintains a view downgrading the importance of competition in nature" (p. 590). It is certainly true that I criticized the quality of the existing evidence concerning the coevolution of competitors. However, my intent was neither to downgrade nor upgrade the importance of competition in nature, but simply to evaluate the evidence for it.

Roughgarden (p. 593) asserts that I maintain a view "biased against the existence of competition." Here he is confusing a specific bias with a general skepticism; I am skeptical of the existence of anything until there is evidence of it. However, it is difficult to counter an allegation of bias, since as Salt (1983) points out, unconscious bias is almost unavoidable. In his autobiography, Charles Darwin said, "I had, also, during many years followed a golden rule, namely, that whenever a published fact, a new observation or thought came across me, which was opposed to my general results, to make a memorandum of it without fail and at once; for I had found by experience that such facts and thoughts were far more apt to escape from the memory than favourable ones" (Darwin 1898, p. 71). The best one can do is to follow Darwin's golden rule and take steps to guard against unconscious bias. I tried to do this in the literature survey in this present paper by strictly defining the criteria for inclusion of studies and, instead of using my own bibliography of references, started from scratch and took the sample from a particular set of journals over a certain span of time.

Since I wrote my 1980 paper I have found two examples of the use of field experiments to investigate the evolution of competitors. Turkington and Harper (1979) experimented with clones of white clover that had coexisted with four different grass species in different sites in a permanent pasture. They transplanted cuttings of each clone to the four different field sites and found that usually transplants grew significantly better in the site in which each had originally lived, both in the presence and absence of the competing grass. Thus local differentiation had apparently occurred in response to both the physical nature of the site and to the species of local competitor. In the greenhouse the cuttings were grown with each of the four grass species using the same soil in all treatments so that only the species of competitor was varied. In three of the four treatments, the clone grew best with the species of grass it had lived with in the field. While it has not yet been established that this differentiation has a genetic basis, it seems likely, given other studies of white clover (references in Turkington and Harper 1979).

The second study (Hairston 1980a) is an extension of the work on competition in salamanders cited earlier (Hairston 1980b, study no. 24). Having established that interspecific competition was less strong in the site where altitudinal overlap was greater, at both sites Hairston replaced local populations of one species with transplants of the same species from the other site. The hypothesis was that where overlap was greater, adaptations that reduced the intensity of competition had been evolved to a greater extent than where overlap was small. The experimental results were consistent with this hypothesis. As in the paper by Turkington and Harper (1979) it has not yet been established that there is a genetic basis for this difference. However, these two papers provide excellent examples of well-designed field experiments that give evidence from natural populations concerning the evolution of competitors.

My views concerning competition have undergone a sea change over the past 20 yr. My first marine research indicated that interspecific competition was an important process in reducing the abundance and reproductive output of the barnacle *Chthamalus* in Scotland (Connell 1961). When I began to study what appeared at first to be a very similar situation in Washington, I was convinced that I would find competition between the barnacles to be of similar importance there. But to my surprise I did not; it was prevented by very intense predation (Connell 1970, 1971, 1975). In both studies, competition was neither upgraded nor downgraded (see Roughgarden 1983), it was simply evaluated along with other interactions in a particular community. Clearly we should withhold judgment until the evidence is more complete. The purpose of the present paper and the experiments suggested in Connell (1980) is to help make the evidence concerning competition both more complete and more rigorous.

LITERATURE CITED

Abramsky, Z., M. I. Dyer, and P. D. Harrison. 1979. Competition among small mammals in experimentally perturbed areas of the shortgrass prairie. Ecology 60:530–536.

Abramsky, Z., and C. Sellah. 1982. Competition and the role of habitat selection in *Gerbillus allenbyi* and *Meriones tristrami:* a removal experiment. Ecology 63:1242–1247.

Adams, E. S., and J. F. A. Traniello. 1981. Chemical interference competition by *Monomorium minimum* (Hymenoptera: Formicidae). Oecologia (Berl.) 51:265–270.

Allen, E. B., and R. T. T. Forman. 1976. Plant species removal and old-field community structure and stability. Ecology 57:1233–1243.

Bertness, M. D. 1981. Competitive dynamics of a tropical hermit crab assemblage. Ecology 62:751–761.
Birch, L. C. 1957. The meanings of competition. Am. Nat. 91:5–18.
Black, R. 1979. Competition between intertidal limpets: an intrusive niche on a steep resource gradient. J. Anim. Ecol. 48:401–411.
Brown, K. M. 1982. Resource overlap and competition in pond snails: an experimental analysis. Ecology 63:412–422.
Burkholder, P. R. 1952. Cooperation and conflict among primitive organisms. Am. Sci. 40:601–631.
Buss, L. W. 1980. Competitive intransitivity and size frequency distributions of interacting populations. Proc. Natl. Acad. Sci. USA 77:5355–5359.
Chappell, M. A. 1978. Behavioral factors in the altitudinal zonation of chipmunks (*Eutamias*). Ecology 59:565–579.
Connell, J. H. 1961. The influence of interspecific competition and other factors on the distribution of the barnacle *Chthamalus stellatus*. Ecology 42:710–723.
———. 1970. A predator-prey system in the marine intertidal region. I. *Balanus glandula* and several predatory species of *Thais*. Ecol. Monogr. 40:49–78.
———. 1971. On the role of natural enemies in preventing competitive exclusion in some marine animals and in rain forest trees. Pages 298–312 *in* P. J. den Boer and G. R. Gradwell, eds. Dynamics of numbers in populations. Proceedings of the Advanced Study Institute on dynamics of numbers in populations, Oosterbeek, 1970. Centre for Agricultural Publishing and Documentation, Wageningen.
———. 1974. Field experiments in marine ecology. Pages 21–54 *in* R. Mariscal, ed. Experimental marine biology. Academic Press, New York.
———. 1975. Some mechanisms producing structure in natural communities: a model and evidence from field experiments. Pages 460–490 *in* M. Cody and J. Diamond, eds. Ecology and evolution of communities. Harvard University Press, Cambridge, Mass.
———. 1976. Competitive interactions and the species diversity of corals. Pages 51–58 *in* G. Mackie, ed. Coelenterate ecology and behavior. Plenum, New York.
———. 1980. Diversity and the coevolution of competitors, or the ghost of competition past. Oikos 35:131–138.
Connell, J. H., and M. J. Keough. 1983. Disturbance and patch dynamics of subtidal marine animals on hard substrata. *In* S. T. A. Pickett and P. S. White, eds. Natural disturbance: an evolutionary perspective. Academic Press, New York (in press).
Creese, R. G., and A. J. Underwood. 1982. Analysis of inter- and intraspecific competition amongst intertidal limpets with different methods of feeding. Oecologia 53:337–346.
Darwin, F., ed. 1898. The life and letters of Charles Darwin. Vol. I. D. Appleton, New York.
Dayton, P. K. 1975. Experimental evaluation of ecological dominance in a rocky intertidal algal community. Ecol. Monogr. 45:137–159.
DeBenedictis, P. A. 1974. Interspecific competition between tadpoles of *Rana pipiens* and *Rana sylvatica:* an experimental field study. Ecol. Monogr. 44:129–141.
Dhondt, A. A., and R. Eyckerman. 1980. Competition between the great tit and the blue tit outside the breeding season in field experiments. Ecology 61:1291–1296.
Dodson, S. I. 1970. Complementary feeding niches sustained by size selective predation. Limnol. Oceanogr. 15:131–147.
Duggins, D. O. 1980. Kelp beds and sea otters: an experimental approach. Ecology 61:447–453.
———. 1981. Interspecific facilitation in a guild of benthic marine herbivores. Oecologia 48:157–163.
Dunham, A. E. 1980. An experimental study of interspecific competition between the iguanid lizards *Sceloporus merriami* and *Urosaurus ornatus*. Ecol. Monogr. 50:309–330.
Dunham, A. E., G. R. Smith, and J. N. Taylor. 1979. Evidence for ecological character displacement in western American catostomid fishes. Evolution 33:877–896.
Eriksson, M. O. G. 1979. Competition between freshwater fish and goldeneyes *Bucephala clangula* (L.) for common prey. Oecologia (Berl.) 41:99–107.
Fonteyn, P. J., and B. E. Mahall. 1981. An experimental analysis of structure in a desert plant community. J. Ecol. 69:883–896.
Fowler, N. 1981. Competition and coexistence in a North Carolina grassland. II. The effects of the experimental removal of species. J. Ecol. 69:843–854.

Friedman, J., and G. Orshan. 1974. Allopatric distribution of two varieties of *Medicago laciniata* (L.) Mill. in the Negev desert. J. Ecol. 62:107–114.

Friedman, J., G. Orshan, and Y. Ziger-Cfir. 1977. Suppression of annuals by *Artemisia herba-alba* in the Negev desert of Israel. J. Ecol. 65:413–426.

Gill, D. E. 1972. Intrinsic rates of increase, saturation densities, and competitive ability. I. An experiment with *Paramecium*. Am. Nat. 106:461–471.

Grace, J. B., and R. G. Wetzel. 1981. Habitat partitioning and competitive displacement in cattails (*Typha*): experimental field studies. Am. Nat. 118:463–474.

Grant, P. 1972. Convergent and divergent character displacement. Biol. J. Linn. Soc. 4:39–68.

———. 1975. The classical case of character displacement. Evol. Biol. 8:237–337.

Gross, K. L. 1980. Colonization by *Verbascum thapsus* (Mullein) of an old field in Michigan: experiments on the effects of vegetation. J. Ecol. 68:919–927.

Gross, K. L., and P. A. Werner. 1982. Colonizing abilities of "biennial" plant species in relation to ground cover: implications for their distribution in a successional sere. Ecology 63:921–931.

Hairston, N. G. 1980a. Evolution under interspecific competition: field experiments on terrestrial salamanders. Evolution 34:409–420.

———. 1980b. The experimental test of an analysis of field distributions: competition in terrestrial salamanders. Ecology 61:817–826.

———. 1981. An experimental test of a guild: salamander competition. Ecology 62:65–72.

Hairston, N. G., F. E. Smith, and L. B. Slobodkin. 1960. Community structure, population control and competition. Am. Nat. 94:421–425.

Harger, J. R. E. 1970a. Comparisons among growth characteristics of two species of sea mussel, *Mytilus edulis* and *Mytilus californianus*. Veliger 13:44–56.

———. 1970b. The effect of species composition on the survival of mixed populations of the sea mussels *Mytilus californianus* and *Mytilus edulis*. Veliger 13:147–152.

———. 1972. Competitive co-existence: maintenance of interacting associations of the sea mussels *Mytilus edulis* and *Mytilus californianus*. Veliger 14:387–410.

Harper, J. L. 1969. The role of predation in vegetational diversity. Brookhaven Symp. Biol. 22:48–62.

———. 1977. Population biology of plants. Academic Press, London.

Hils, M. H., and J. L. Vankat. 1982. Species removals from a first-year old-field plant community. Ecology 63:705–711.

Hixon, M. A. 1980. Competitive interactions between California reef fishes of the genus *Embiotoca*. Ecology 61:918–931.

Holbrook, S. J. 1979. Habitat utilization, competitive interactions, and coexistence of three species of cricetine rodents in east-central Arizona. Ecology 60:758–769.

Hurlberg, L. W., and J. S. Oliver. 1980. Caging manipulations in marine soft-bottom communities: importance of animal interactions on sedimentary habitat modifications. Can. J. Fish. Aquat. Sci. 37:1130–1139.

Inouye, D. W. 1978. Resource partitioning in bumblebees: experimental studies of foraging behavior. Ecology 59:672–678.

Inouye, R. S. 1981. Interactions among unrelated species: granivorous rodents, a parasitic fungus and a shared prey species. Oecologia 49:425–427.

Jackson, J. B. C. 1979. Overgrowth competition between encrusting ectoprocts in a Jamaican cryptic reef environment. J. Anim. Ecol. 48:805–824.

———. 1981. Interspecific competition and species distributions: the ghosts of theories and data past. Am. Zool. 21:889–901.

Kastendiek, J. 1982. Factors determining the distribution of the sea pansy, *Renilla kollikeri*, in a subtidal sand-bottom habitat. Oecologia 52:340–347.

Kay, A. M., and M. J. Keough. 1981. Occupation of patches in the epifaunal communities on pier pilings and the bivalve *Pinna bicolor* at Edithburgh, South Australia. Oecologia 48:123–130.

Keough, M. J. 1983. Effects of patch size on the abundance of sessile marine invertebrates. Ecology (in press).

Kroh, G. C., and S. N. Stephenson. 1980. Effects of diversity and pattern on relative yields of four Michigan first year fallow field plant species. Oecologia 45:366–371.

Larson, R. J. 1980. Competition, habitat selection, and the bathymetric segregation of two rockfish (*Sebastes*) species. Ecol. Monogr. 50:221–239.

Lawton, J. H., and M. P. Hassell. 1981. Asymmetrical competition in insects. Nature 289:793–795.
Levine, S. H. 1976. Competitive interactions in ecosystems. Am. Nat. 110:903–910.
Lubchenco, J. 1980. Algal zonation in the New England rocky intertidal community: an experimental analysis. Ecology 61:333–344.
Lynch, M. 1978. Complex interactions between natural coexploiters—*Daphnia* and *Ceriodaphnia*. Ecology 59:552–564.
McClure, M. S., and P. W. Price. 1975. Competition among sympatric *Erythroneura* leafhoppers (Homoptera: Cicadellidae) on American sycamore. Ecology 56:1388–1397.
Mackie, G. L., S. U. Qadri, and R. M. Reed. 1978. Significance of litter size in *Musculium securis* (Bivalvia: Sphaeriidae). Ecology 59:1069–1074.
McLay, C. L. 1974. The distribution of duckweed *Lemna perpusilla* in a small southern California lake: an experimental approach. Ecology 55:262–276.
Menge, B. A. 1976. Organization of the New England rocky intertidal community: role of predation, competition and environmental heterogeneity. Ecol. Monogr. 46:355–393.
Menge, B. A., and J. P. Sutherland. 1976. Species diversity gradients: synthesis of the roles of predation, competition and temporal heterogeneity. Am. Nat. 110:351–369.
Miles, J. 1974. Effects of experimental interference with stand structure on establishment of seedlings in callunetum. J. Ecol. 62:675–687.
Minot, E. O. 1981. Effects of interspecific competition for food in breeding blue and great tits. J. Anim. Ecol. 50:375–385.
Montgomery, W. I. 1981. A removal experiment with sympatric populations of *Apodemus sylvaticus* (L.) and *A. flavicollis* (Melchior) (Rodentia: Muridae). Oecologia (Berl.) 51:123–132.
Morse, D. H. 1981. Interactions among syrphid flies and bumblebees on flowers. Ecology 62:81–88.
Neill, W. E. 1974. The community matrix and interdependence of the competition coefficients. Am. Nat. 108:399–408.
Pacala, S., and J. Roughgarden. 1982. Resource partitioning and interspecific competition in two two-species insular *Anolis* lizard communities. Science 217:444–446.
Paine, R. T., and R. L. Vadas. 1969. The effects of grazing by sea urchins, *Strongylocentrotus* spp., on benthic algal populations. Limnol. Oceanogr. 14:710–719.
Peterson, C. H. 1979. The importance of predation and competition in organizing the intertidal epifaunal communities of Barnegat Inlet, New Jersey. Oecologia 39:1–24.
———. 1982. The importance of predation and intra- and interspecific competition in the population biology of two infaunal suspension-feeding bivalves, *Protothaca staminea* and *Chione undatella*. Ecol. Monogr. 52:437–475.
Peterson, C. H., and S. V. Andre. 1980. An experimental analysis of interspecific competition among marine filter feeders in a soft-sediment environment. Ecology 61:129–139.
Petranka, J. W., and J. K. McPherson. 1979. The role of *Rhus copallina* in the dynamics of the forest-prairie ecotone in north-central Oklahoma. Ecology 60:956–965.
Pinder, J. E. 1975. Effects of species removal on an old-field plant community. Ecology 56:747–751.
Price, M. V. 1978. The role of microhabitat in structuring desert rodent communities. Ecology 59:910–921.
Quinn, J. F., and A. E. Dunham. 1983. On hypothesis testing in ecology and evolution. Am. Nat. 122:602–617.
Race, M. S. 1982. Competitive displacement and predation between introduced and native mud snails. Oecologia 54:337–347.
Rahman, M. S. 1976. A comparison of the ecology of *Deschampsia cespitosa* (L.) Beauv. and *Dactylis glomerata* (L.) in relation to the water factor. I. Studies in field conditions. J. Ecol. 64:449–462.
Raynal, D. J., and F. A. Bazzaz. 1975. Interference of winter annuals with *Ambrosia artemisiifolia* in early successional fields. Ecology 56:35–49.
Redfield, J. A., C. J. Krebs, and M. J. Taitt. 1977. Competition between *Peromyscus maniculatus* and *Microtus townsendii* in grasslands of coastal British Columbia. J. Anim. Ecol. 46:607–616.
Robertson, D. R., H. P. A. Sweatman, E. A. Fletcher, and M. G. Cleland. 1976. Schooling as a mechanism for circumventing the territoriality of competitors. Ecology 57:1208–1220.
Roughgarden, J. 1983. Competition and theory in community ecology. Am. Nat. 122:583–601.

Rubin, J. A. 1982. The degree of intransitivity and its measurement in an assemblage of encrusting cheiolostome bryozoans. J. Exp. Mar. Biol. Ecol. 60:119–128.
Russ, G. R. 1982. Overgrowth in a marine epifaunal community: competitive hierarchies and competitive networks. Oecologia 53:12–19.
Salt, G. W. 1983. Roles: their limits and responsibilities in ecological and evolutionary research. Am. Nat. 122:697–705.
Schoener, T. W. 1974. Resource partitioning in ecological communities. Science 185:27–39.
———. 1983. Field experiments on interspecific competition. Am. Nat. 122:240–285.
Schroder, G. D., and M. L. Rosenzweig. 1975. Perturbation analysis of competition and overlap in habitat utilization between *Dipodomys ordii* and *Dipodomys merriami*. Oecologia 19:9–28.
Seifert, R. P., and F. H. Seifert. 1976. A community matrix analysis of *Heliconia* insect communities. Am. Nat. 110:461–483.
———. 1979. A *Heliconia* insect community in a Venezuelan cloud forest. Ecology 60:462–467.
Simberloff, D. 1983. Competition theory, hypothesis-testing, and other community ecological buzzwords. Am. Nat. 122:626–635.
Slobodkin, L. B., F. E. Smith, and N. G. Hairston. 1967. Regulation in terrestrial ecosystems, and the implied balance of nature. Am. Nat. 101:109–124.
Smith, D. C. 1981. Competitive interactions of the striped plateau lizard (*Sceloporus virgatus*) and the tree lizard (*Urosaurus ornatus*). Ecology 62:679–687.
Smith, D. W., and S. D. Cooper. 1982. Competition among *Cladocera*. Ecology 63:1004–1015.
Smith-Gill, S. J., and D. E. Gill. 1978. Curvilinearities in the competition equations: an experiment with ranid tadpoles. Am. Nat. 112:557–570.
Sousa, W. P. 1979. Experimental investigations of disturbance and ecological succession in a rocky intertidal algal community. Ecol. Monogr. 49:227–254.
Strong, D. R. 1982. Harmonious coexistence of hispine beetles on *Heliconia* in experimental and natural communities. Ecology 63:1039–1049.
Sutherland, J. P. 1978. Functional roles of *Schizoporella* and *Styela* in the fouling community at Beaufort, North Carolina. Ecology 59:257–264.
Taylor, P. R., and M. M. Littler. 1982. The roles of compensatory mortality, physical disturbance, and substrate retention in the development and organization of a sand-influenced, rocky-intertidal community. Ecology 63:135–146.
Tinkle, D. W. 1982. Results of experimental density manipulation in an Arizona lizard community. Ecology 63:57–65.
Turkington, R., M. A. Cahn, A. Vardy, and J. L. Harper. 1979. The growth, distribution and neighbour relationships of *Trifolium repens* in a permanent pasture. III. The establishment and growth of *Trifolium repens* in natural and perturbed sites. J. Ecol. 67:231–243.
Turkington, R., and J. L. Harper. 1979. The growth, distribution and neighbour relationships of *Trifolium repens* in a permanent pasture. IV. Fine-scale biotic differentiation. J. Ecol. 67:245–254.
Underwood, A. J. 1978. An experimental evaluation of competition between three species of intertidal prosobranch gastropods. Oecologia 33:185–202.
Underwood, A. J., and E. J. Denley. 1984. Paradigms, explanations and generalizations in models for the structure of intertidal communities on rocky shores. *In* D. R. Strong, Jr., D. Simberloff, L. G. Abele, and A. B. Thistle, eds. Ecological communities: conceptual issues and the evidence. Princeton University Press, Princeton, N.J.
Van der Plank, J. E. 1968. Disease resistance in plants. Academic Press, New York.
Vandermeer, J. 1980. Indirect mutualism: variations on a theme by Stephen Levine. Am. Nat. 116:441–448.
Werner, E. E., and D. J. Hall. 1977. Competition and habitat shift in two sunfishes (Centrarchidae). Ecology 58:869–876.
Werner, P. A. 1977. Colonization success of a "biennial" plant species: experimental field studies of species cohabitation and replacement. Ecology 58:840–849.
Wiens, J. A. 1977. On competition and variable environments. Am. Sci. 65:590–597.
Wilbur, H. M. 1972. Competition, predation and the structure of the *Ambystoma-Rana sylvatica* community. Ecology 53:3–21.

Williams, A. H. 1981. An analysis of competitive interactions in a patchy back-reef environment. Ecology 62:1107–1120.
Wise, D. H. 1981a. Inter- and intraspecific effects of two orb-weaving spiders (Araneae: Araneidae). Oecologia (Berl.) 48:252–256.
———. 1981b. A removal experiment with darkling beetles: lack of evidence for interspecific competition. Ecology 62:727–738.
Wit, C. T. de 1960. On competition. Versl. Landbouwk. Onderz. (Agric. Res. Rep.) no. 66, 8. Wageningen, Netherlands.
Woodin, S. A. 1974. Polychaete abundance patterns in a marine soft-sediment environment: the importance of biological interactions. Ecol. Monogr. 44:171–181.

ROLES: THEIR LIMITS AND RESPONSIBILITIES IN ECOLOGICAL AND EVOLUTIONARY RESEARCH

George W. Salt

Department of Zoology, University of California, Davis, California 95616

> The mind lingers with pleasure upon the facts that fall happily into the embrace of the theory, and feels a natural coldness toward those that assume a refractory attitude [Chamberlin (1897, p. 840)].

In any dispute between A and B, there are four possible outcomes: A can be right and B wrong; B right and A wrong; both can be wrong; or both can be right. The first two results are familiar from athletic contests and legal disputes, while those of the third kind can be readily experienced at bars and cocktail parties. Seemingly, the fourth class of disputes should not occur at all. In the past, however, some heated scientific debates have been resolved by the finding that both disputants were correct and that the argument arose because they were considering different facets of the same problem. It will be the thesis presented in the following pages that the seeming disagreement between the participants in this round table fall into this fourth category.

It would, of course, be possible to discuss these articles in terms of their differences, to judge which points were well taken and which were not, and so forth. To do so would focus attention on a secondary and probably transient value of the round table. The major value is that when all the papers are considered together, one can discern the outlines of an evolving procedure for conducting research in ecology and evolutionary biology. The character of various research roles is being made apparent, and with them, the functions for which each is responsible. Each of the authors in the round table is, in fact, helping to crystallize the dimensions of these roles and functions.

A scientific discipline is recognized by its body of laws, principles or generalizations which apply widely to seemingly diverse systems and which permit more or less successful predictions of performance of these systems under different circumstances. Both ecologists and evolutionary biologists search for such basic findings in their respective fields. The manner in which this search should be conducted is rarely explicitly discussed, but for the first half of this century, almost by consensus, the only defensible way in which to conduct inquiry was through induction. True, there were early experimental studies in plant ecology by Watt and others (see Jackson 1981) and by Park and Gause in animal ecology, and a variety of physiological studies by both plant and animal biologists, but the major thrust of investigation was in the accumulation, collation, and recording of observations. The means by which generalizations were to result from these endeavors was never consciously discussed, so that the term induction as used here does not meet the standards of rigor which would be demanded by a philosopher of science. Calow and Townsend (1981) have typified this method as the a posteriori approach. In mid-twentieth century, with the appearance of high speed digital computers, it appeared that induction could be vastly simplified and that major conclusions would emerge from massive data analysis. By and large, it did not happen.

At the same time, a variety of individuals began to express dissatisfaction with induction. Its defects had become evident. While there was no doubt that it could produce major generalizations when used by a perceptive intellect, it was slow and tended to be ponderous and dull. More importantly, because it required only a most general focus, or no focus at all, it encouraged all manner of data collection and reporting. Masses of information were accumulated with the expectation that at some point a critical density would be reached and generalizations would precipitate out spontaneously. The cost, in money and research time, was large and the results were small.

Two alternatives to the inductive method appeared. Verhulst, Lotka, and Volterra provided the basis for subsequent developments in mathematical treatments of intraspecific population dynamics by Cole, Nicholson, Birch, and others. A similar line of development occurred in population genetics. Both these lines of research continue today. The efforts of these individuals to provide models or generalizations about interspecific interactions have been less successful.

For interspecific interactions, the alternative provided by those unhappy with induction was the hypothetico-deductive method. Its principal advocates in the United States were G. E. Hutchinson, Robert MacArthur, and a group of like-minded individuals. Calow and Townsend (1981) have referred to it as the a priori method. It seemed to answer much of the criticisms against induction. One inferred what seemed to be the general features from a system and made a hypothesis about how these features interacted. The hypothesis was then cast into a mathematical statement. A series of manipulations on the resulting equation produced a variety of hitherto unappreciated relationships within the system. The legitimacy of the procedure was determined by two things: the accuracy of the initial hypothesis and the correctness of the subsequent manipulations of the equations. Verification of the predictions usually took the form of looking for a rough concordance with field observations, often as not on the same system which had prompted the effort in the first place.

The advent of this method had a revivifying effect on ecology and evolutionary biology. First, since it considered only the most general aspects of systems and measured results only in comparative terms, such as larger or smaller or steeper or shallower, it appeared to produce generalizations in abundance. Second, it provided field biologists and experimentalists with specific topics for investigation. Research became focused rather than diffuse and idiosyncratic. The entire intellectual climate in ecology and evolutionary biology became more taut and incisive. While suggestions made by the hypothetico-deductionists were not universally accepted, debate and discussion in ecology and evolutionary biology came to be focused on concepts and proposed explanations and mechanisms. Those using the hypothetico-deductive method felt willing to consider large and important topics. They concerned themselves with the mechanisms determining species composition of communities, the relationship between community complexity and stability, and the determinants of species compositions of island faunas. All this promised a union between ecologists concerned with adaptation and geneticists studying population genetics. A unified theory of evolution based on competition and adaptation seemed to be possible.

Unfortunately, as time passed, the weaknesses of the hypothetico-deductive method became apparent. It was simplistic. Great leaps over masses of relationships were necessary in order to get from the initial hypotheses to the final conclusions. Data accumulation and analysis seemed to be distained. Its hypotheses were deterministic, and it was divisive.

As the mathematics used became more esoteric, the field biologists found themselves either unwilling or unable to keep abreast while at the same time maintaining their necessary tools in systematics, statistics, physiology, and morphology. They also resented the seeming arrogation of the generalist title by the mathematically inclined. The fraternity became splintered into factions with lofty or derogatory titles depending on who applied them. "Empiricists" and "theoreticians" were the most polite.

Secondly, if one of the failings of induction was that it fostered and legitimatized much al fresco hack work, hypothetico-deduction was no less imperfect. It stimulated ever more delicate and elegant manipulations on more and more extravagantly generalized model systems so that the specter of the rebirth of scholasticism began to appear.

Many of the theoreticians' generalizations, whether called theories, principles, or whatever, were tested and found, not surprisingly, to be either incorrect or inadequate. They were presumably intended to be considered no more than reasonable suggestions. Nonetheless, the theoreticians felt somewhat beleaguered as a result, while still maintaining that their method had merit. Many empiricists, meanwhile, were approaching a frustrated state. Consciously or unconsciously, they recognized how stimulating the results of modeling had been and came, albeit reluctantly, to value the contributions of the theoreticians. However, thanks to the dubious but nonetheless popular cachet of legitimacy provided by mathematics to an idea, a theoretician's hypotheses were likely to be accepted until demonstrated false. Because empirical tests of hypotheses are time consuming, the empiricists could contemplate the prospect of an ever increasing array of hypotheses, most of which were probably incorrect, but which would probably be accepted until so demonstrated. If every mathematically generated hypothesis had to be tested empirically, they would never keep up.

This narrative, then, is a view of the character of the intellectual climate in ecology and evolutionary biology until recent times. It does not take into account much substantial progress made outside these two intellectual streams: the experimental and mathematical studies on predation and insect parasitism by Varley, Hassell, Holling, Paine, and Murdoch, for example. The aim has been to describe the interaction between the major ideas controlling the manner in which research was designed and executed.

At this point, we can fairly say that neither induction nor hypothetico-deduction enjoys the wholehearted support of research investigators in ecology and evolution. The time is ripe to move on to a new approach which combines the strengths of both methods and minimizes the deficiencies. One can discern in the writings of the authors in this round table the emergence of a new format for the conduct of research in these two related fields. Roles are being defined and the limits and responsibilities of them can be seen.

Polya (1957) in his description of the heuristic method, declared that a person seeking to solve a mathematical problem had two tasks before him: "a solution to find" and "a solution to prove." He thus defined two functions which were involved in problem solving. The person performing a function was playing a particular role. We may adopt this point of view with respect to problem solving in ecology and evolutionary biology by expanding the number of functions required and by pointing out that one individual may play more than one role. Further, as will be discussed later, each role has certain limits and responsibilities. Translating Polya's language to apply to ecological and evolutionary biology, those seeking explanations for problems have to: (1) find an explanation for the phenomenon and then (2) demonstrate that it is the correct one. The first role is that of an observer. An observer studying a relationship, performance, or characteristic of populations or communities in the field nearly always generates explanations for the problem he or she is studying. This explanation is usually in the form of a verbal model. Sometimes it may be an unconscious one. If it proves to be acceptable or ultimately proves correct, it represents the use of the inductive method at its best. It worked this way for Darwin, Elton, and a variety of other investigators. The explanation should properly be considered a hypothesis under the tenets being advocated here.

If, however, the relationships being observed are complex, the formulation of a verbal model becomes difficult if not impossible. Even if one can be made, it does not lend itself to further manipulation and examination for hidden relationships. Often a verbal model cannot be constructed. The observer is overwhelmed by facts, data, exceptions, aberrant readings, and a general surfeit of material to the point that no single or simple explanation will fit all the information at his command. The currently popular response is to subject the data pile to one of the various techniques of multivariate analysis in the hope that, in the process, something of value will be sifted out.

The second role is that of the modeler or theoretician, who, by contrast, specializes in abstracting only certain elements from a complex of information. Presumably he or she uses his or her own biological insight in making the choice, but can make repeated selections and models using different elements. Then, through a series of manipulations, an array of predicted relationships or performances are produced, a hypothesis about the system in question. Theorizing of this kind stands or falls on three criteria: the appropriateness of the starting assumptions or simplifications; the relationships specified by the structure of the model; and the worth of the predictions produced.

The observer and the modeler, thus are working in parallel to produce hypotheses ("solutions to find") about the problem. For complex problems, the observers probably need the abstracting abilities of the modelers for the generation of potential explanations of their studied phenomena; and, the modelers need the observers for the delineation of the systems they will model. Considering how complementary their roles are, it is bewildering that rancorous relationships sometimes exist between them.

The hypotheses or explanations must be tested ("a solution to prove"). In these times, this task falls more and more to the experimentalist, the third role. Given a suitable series of experimental tests, the investigator can identify which of the proffered explanations, or none of them, is the most likely to be correct. If all goes well, one hypothesis will be found acceptable as a provisional solution.

Although this step would complete the process under Polya's scheme, there is still one further function needed. Decisive experimental programs are difficult to design and even more difficult to execute. There is a worrisome sanctity developing about experimental results. The fact is, because of faulty or inappropriate design and execution, a great many experiments prove nothing except that the individual attempted an experiment. A historical example will illustrate.

Gause (1934) conducted an experimental test of the Lotka-Volterra predator-prey model. The model specifies a prey population which grows exponentially and a predator population whose reproduction is a function of the number of prey consumed. For his test, Gause used *Paramecium caudatum* which grows logistically and *Didinium nasutum* whose reproductive rate is independent of the rate of food intake. With keen hindsight, we can see that as a test of the model, the experiments were meaningless. In effect, what Gause did was test a model which had not been formulated. This incongruence was never noted, with the result that his experiments are cited relative to the Lotka-Volterra model in textbooks to this day.

This episode points to the necessity for the performance of a fourth role in the process, a judge of the experiment who decides whether, in fact, it demonstrates anything or not. Have the hypotheses which were specified as incorrect by the experiment actually been falsified or not? Indeed, the role of the judge is required for each of the other two functions in the research sequence. At present, the judges' roles are largely filled by reviewers of manuscripts before publication, but not always in a rigorous fashion. A review of a theoretical paper will often contain the disclaimer, "I have not worked through the math, but the conclusions look all right to me." What is needed is a formal recognition of this office, both by the ecological and evolutionary biology community and by editors.

The question now arises: Who is to fill the roles? There is a philosophy which uses the mythical "Renaissance Man" as its model. Adherents to these beliefs would have it that a good investigator should assume all roles: Make his own observations, devise a model and generate hypotheses to explain the results, conduct experiments to test his hypotheses, and act as judge all through the process. Assuming an investigator possessed this array of talents, such a procedure would be unsound and unconvincing. The problem is unconscious bias.

As Chamberlin (1897) pointed out many years ago, it is illogical to expect an individual to follow the classical protocol for experimentation in which one generates a hypothesis and then attempts to disprove it. It asks that the individual commit infanticide on his or her own brainchild. The result is a powerful impulse toward unconscious bias in the conduct of the experiment and the interpretation of the results.

On the other hand, ecologists and evolutionary biologists are not unidimensional individuals. Most of them would not be content to confine their activities to one role. The way out of this dilemma seems to be to postulate that an individual should not play more than one role on one piece of research (leaving the word "piece" purposely vague).

An observer is likely to value most highly those masses of data which were most difficult to obtain or are most unique. Were he or she capable of constructing a model of his/her information, the likelihood is that these valuable observations would occupy a central place. Indeed, he/she would probably be loath to ignore any of the information so carefully gathered. The resulting model, thus biased, would probably be cumbersome at best and very likely intractable, to use one of the modelers' favorite words.

The modeler or theoretician has a special problem. As will be developed further on, it is possible for this group to generate more models than the experimentalists can ever test. As a consequence, only a portion of the models will be included in the research process sketched here. Only those which appear to be most cogent and promising will be tested. It is obviously in the modeler's interest to place his/her creation in this favored category. Consequently, one must make a distinction between substantiation of a model and its being tested.

The difference can best be illustrated by analogizing the modeler with a manufacturer of medicines. The manufacturer will make powerful efforts to ensure that the formulation of the compound is carefully and correctly done just as a modeler will do with a model. The manufacturer will also make elaborate studies to verify that the proposed medicine performs as it is supposed to. A modeler can make similar investigations both of the assumptions and the predictions of the model. However elaborate the pharmaceutical house may be in its studies, the product will not be certified for public use without a test being performed on it by a disinterested person or laboratory, usually in a double blind experiment.

The predictions of an ecological or evolutionary model require the same kind of test. This conclusion may seem unnecessarily stringent until one remembers that currently models are being advanced as a basis for decisions on the size of natural areas, refuges, and National Parks, for example. A modeler can carry out extensive studies to help substantiate the validity of the model, but the definitive test should be made by another individual who is without a personal stake in the outcome.

The common conception is that experimentalists are unbiased, but the fact that they are the very ones who invented blind and double-blind procedures argues against this idea. Such elaborate precautions are not required except in issues involving human welfare, but the bases for the development of such techniques need to be remembered by all experimentalists and opportunities for the expression of bias in their work minimized.

Lastly, there is the matter of judging. An individual cannot, unless blessed with almost total psychological security and a private income, act effectively as a judge of his or her own work. The pressures for recognition and publication are too strong. One need only to inspect the rejection rate from journals to see this conclusion verified.

We need recognition of the role of critic or judge as an independent activity. Other fields have just such functions specified as part of their professional body: art critics, legal judges, and so on, but for some reason this office has never received formal identification in scientific communities. The presumption appears to be that all individuals are acting as judges at all times. This view appears to be too optimistic. It is a difficult task to be both well informed, even-handed, and decisive at the same time. Authors of review articles have the opportunity to act as judges. To do so, however, requires that they be demanding in their standards when they survey the literature on a topic, identifying those studies which are meaningful and those which are not. It would help popularize this activity if those judging could stifle their impulses toward displays of penetrating wit at the expense of the persons being judged. Similarly, if the judging process were more formally recognized and executed, the feelings of being singled out for persecution which it generates would be lessened. The irritation could be reduced further if all judging were scrupulously fair and appropriate. Unconscious bias is probably more destructive when it appears in judging than in any other place in the research process.

On a technical level, judging can be done laterally. Observers judge the validity of the observations. Was it wise to use a plankton net instead of a water bottle? Are strip transects appropriate for the kinds of counts made? Does a light trap sample the area in the manner stated? And so on. Similar remarks could be fashioned for the other roles. At a more meaningful level, major judgments are made by those occupying the next step in the sequence. Observations which are not sufficiently interesting to stimulate the efforts of a theoretician have, in effect, been judged. A model which generates no independent test of it has been ignored.

Judgments made in the reverse direction, however, are probably not dependable. An observer of a phenomenon is not a good judge of a model related to those observations. Similarly, the modeler is likely to have an unconscious bias when examining the results of an experimental test of his or her hypotheses. Whatever their private opinions, one can suggest that an individual's performance of a role does not include the function of judging the processes and results of the next succeeding step in the research sequence insofar as it involves the results of his/her own investigations.

Those individuals who accept that they are participating in a sequential research process will also recognize that they have certain responsibilities regarding the role they play. An observer will be conscious of the fact that his or her data may form the material on which a theoretician may base a model and take pains to acquaint himself/herself with the procedures and restrictions under which modelers operate. The quality and relevance of a model will be determined to a measurable degree by the quality of the observations on which it is based. Cohen's (1978) stimulating analysis of food webs was less compelling than it might have been. The flaw lay not in his work but in the data compiled by the observers, some of whom simultaneously classified some diet items to species and placed other food items in diffuse categories such as "protozoa," or even worse, "algae."

In much the same manner as observers and experimentalists test methods, theoreticians frequently devote their energies to analyses of other persons' models: converting them from deterministic to stochastic, relaxing one of the assumptions and determining the consequences, making sensitivity analyses, and such like activities. This intellectual interchange undoubtedly has value in developing new structures for models or new techniques of manipulation, but the true clients of the theoreticians are the experimentalists. A modeler who ignores this responsibility has, in effect, placed him- or herself outside the process of research as formulated here. Further, the modeler has the responsibility of formulating results in such a fashion that they can be quantified and approximated in living systems. Postulations about the behavior of fanciful organisms cannot be tested experimentally no matter how much heuristic value is claimed for them.

Similarly, if experimentalists are to judge which models and their predictions are worthy of their attention, then they need to be able to identify the conditions specified in the mathematics of the model if they are to avoid the trap that Gause fell into. They either require sufficient command of mathematics to be able to dissect the model and identify its conditions, or they should solicit the help of another modeler or mathematician who can.

In many versions of the "scientific method," experimentation is the end of the process. The hypothesis which remains unfalsified at the termination of the program is accepted, at least provisionally, as the correct one. This conclusion does not apply in ecology and evolutionary biology. Results from the constrained and manipulated environment of the experiment may not be mirrored in field situations. A preferred food item in an experimental determination may be so rare in the field as to play no part in the organism's diet. It follows, then, that the results of experimental determinations must be verified by the observers, the experimentalist's major clients. It behooves the experimentalist to remember this needed verification of his or her findings when designing the investigation. If the experimental conditions are too rarified or the circumstances are too unnatural, the observers will not be able to apply the findings in natural systems. It is for this reason that field experiments are so highly regarded despite their inherent difficulties in both design and execution.

Given the requirement that an individual be aware of his/her responsibilities to those playing the next role in the sequence and the necessity of being informed about the technicalities of the preceding step in order to make worthy judgments, the "Renaissance Man" begins to look more attractive as an ideal. Were it possible for us all to function in this fashion—observer on this project, modeler of that process, and experimentalist on a third, all the while acting as judge both laterally and anteriorly—there would only be a need to emphasize responsibilities. Even those admonitions would be unnecessary were this interrole transferring the universal manner of operation. To the degree that it is not, the foregoing comments are pertinent.

It may be commented that the roles outlined above are not new. What is being suggested here is that there exists a need for explicit recognition of each and of its limitations and responsibilities. If an individual shifts his or her activities from one role to another, he or she should also recognize the new constraints operating. The individual, for example, who has learned a new technique and sets about an investigation because "I thought of this real neat experiment I could do" is not filling the role of an experimentalist but that of an observer. The work is not intended to test the work of either an observer or a theoretician.

To return to the dispute between A and B with respect to this round table, one can recognize that the fourth outcome applies: They are all right. Each of them is filling one or more roles much in the fashion outlined here. Where there is disagreement, it frequently results from a lack of recognition of the fact that one individual may fill several roles and consequently have several sets of limits and responsibilities. The judging process is also evident although not formally recognized as such.

If it is accepted that the research process consists of sequential steps with the accompanying judgmental processes, then it follows that all the steps are necessary, and no one is more important or vital than any other. The effectiveness and efficiency with which the series operates will depend on the degree to which the roles are accepted. Although authors in this round table deal in specifics, in a larger sense they are engaged in the process of delineating roles, limits, and responsibilities. In so doing, they are defining the manner in which future research should be conducted in ecology and evolutionary biology.

ACKNOWLEDGMENT

It would be gratifying if the ideas sketched above were original, but they probably are not. More likely, they represent an amalgam of concepts and opinions absorbed subliminally from many authors whom I cannot now identify. To those who recognize their thoughts in the pages above, I thank you for your contributions even though I cannot acknowledge them formally.

LITERATURE CITED

Calow, P., and C. R. Townsend. 1981. Introduction. Pages 3–19 *in* C. R. Townsend, and P. Calow, eds. Physiological ecology: an evolutionary approach to resource use. Sinauer, Sunderland, Mass.

Chamberlin, T. C. 1897. The method of multiple working hypotheses. J. Geol. 5:837–848.

Cohen, J. E. 1978. Food webs and niche space. Monogr. Popul. Biol. no. 11. Princeton University Press, Princeton, N.J.

Gause, G. F. 1934. The struggle for existence. Hafner, New York.

Jackson, J. B. C. 1981. Interspecific competition and species distributions: the ghosts of theories and data past. Am. Zool. 21:889–901.

Polya, G. 1957. How to solve it. A new aspect of mathematical method. 2d ed. Princeton University Press, Princeton, N.J.

INDEX

A priori method. *See* Hypothetico-deductive method

Alpha. *See also* Power analysis; Type I error
defined, 38
standard, 39

Alternative hypotheses. *See also* Noninteraction; Null hypotheses
cause of distributional limits, 28
listed vs. competition, 49, 59–60, 62, 63–74
need for, 5–6, 33, 52
nonexclusiveness, 25
null models as, 12
random colonization as, 30–31
testing, 53. *See also* Hypothesis testing

Beta. *See also* Power analysis; Type II error
defined, 38

Bias, 87, 102, 110, 120–122
against competition, 13
in observations categorized by model, 25
toward interference mechanisms, 10
unequal treatment of hypotheses leading to, 18

Body size of lizards, 17, 19, 63

Certainty, 4, 58

Character displacement, 12, 16, 56, 63–65

Chi-square test in power analysis, 39–40

Climate. *See* Weather

Coevolution, 56, 62–63, 109
demonstration of, 5, 8, 13. *See also* Theory, evaluating
systems dimensions effect, 10–11

Common sense, 3, 46–48, 49, 57

Competition. *See also* Interspecific competition; Intraspecific competition
defined, 82

Confirmation, 7, 47. *See also* Falsification

Connell, Joseph, protocol of, 5, 8–11, 53, 109
divergence, 109
systems dimensions, 10–11

Deductive method. *See* Hypothetico-deductive method

Density-vague population ecology, 66–67

Divergence in Connell's protocol, 8, 109

Empirical method. *See also* Experiment; Experimental design
possibility of, 5, 65

Empiricism, 62–63, 119. *See also* Hypothetico-deductive method; Popper, Karl; Strong inference

Experiment. *See also* Field experiments; Transplant/removal experiment
role of, 29, 124

Experimental design, 82–85, 109, 121
controls, 83–84
generalization from results, 84
population density, 83, 84
without models, 17

Exploitative mechanisms, 10, 64

Fact, evaluating claims of. *See* Theory, evaluating

Falsification, 6, 7, 23, 24–25, 47, 48, 49, 52
of probability estimates, 23, 41

Field experiments, 81–103
listed, 104–109, 111
reliance on, 9
without models, 17

Formality, 5, 47. *See also* Connell, Joseph, protocol of; Hypothetico-deductive method; Strong inference
vs. common sense, 3–4, 46–47

Genetic drift, 11

Genetic variation, 110
  in Connell's protocol, 8, 9
  transplant/removal experiment to detect, 9

Hairston et al. hypothesis. *See* HSS hypothesis

Heuristic method, 120

HSS hypothesis
  defined, 89
  tested, 89–92

Hyperparasitism, 70

Hypothesis testing, 53, 58, 22–34, 110, 121
  decision table, 39
  hypothetico-deductive method, 24–26
  statistical, 23, 41. *See also* Power analysis

Hypothetico-deductive method, 24–26, 34, 118–120

Inductive method, 23, 52, 117–120

Interacting hypotheses, 27–29

Interference, 10, 64

Interspecific competition, 63–65, 67–74, 81–83, 85–103. *See also* Niche theory
  acceptance of, 14
  alternatives to, 49, 53, 56, 59–60. *See also* Density-vague population dynamics
  and body size, 19, 63, 92–93
  and coevolution, 5
  and niche theory, 15
  and predator-prey dynamics, 10, 28
  annual variation, 93, 96
  asymmetry, 97–100, 102
  contingency test for, 43
  decomposer insects, 68–69
  frequency, 86–89
  habitat as alternative, 49
  in Connell's protocol, 8–9
  in vulnerable groups, 92–93
  measuring variation, 85
  measurement defined, 86, 93
  noninsect parasites, 71–72
  plants, 72
  predatory and parasitic insects, 69–71
  rank, 97–100, 102
  variation, 93, 95, 96, 109
  vertebrates, 16, 19, 63, 73–74, 93–96

Intraspecific competition, 81–82, 83, 102
  strength vs. interspecific competition, 97, 98

Island studies, 11, 29–31
  as random subsamples, 30

Keystone predator effect, 28

Logical primacy (of null hypothesis), 6, 32, 49, 59. *See also* Temporal primacy

Lotka-Volterra model, 16, 121

Mathematical theory, 62–63, 118, 119, 120, 124
  coevolution, 16, 19. *See also* Lotka-Volterra model

Migration-coupling, 18

Model construction, 14–15, 17, 30–31, 120–121, 122
  as a discipline, 50
  failure to describe system, 16
  limits to realism, 53

Model
  defined, 14
  simplifying, 16
  summarizing, 15–16
  testing, 15–16. *See also* Hypothesis testing

Morphology, 12, 13, 63–65

Multiple causality, 5, 23, 24, 25, 27–29

Natural selection and competition, 1

Natural variability, 62

Neutrality hypothesis (of population genetics), 11–12

Niche theory, 52, 56, 60–62, 67–74, 82
  alternatives to, 62
  as model, 15
  breadth, 9, 94, 110
  coevolution, 5
  divergence, 5
  separation, 12
  shifts, 9, 82, 110

INDEX

Non-alternative hypotheses, 26–29, 33

Noninteraction
  as null hypothesis, 6, 25, 29–32, 49
  estimating parameters for, 31

Null hypotheses, 5–7, 24–25, 29–33, 58–59. *See also* Logical primacy; Noninteraction
  construction problems, 6, 25–26, 29, 59
  in plant succession, 27
  of Connor and Simberloff, 11–13, 19
  of Strong, 6, 11–13, 19
  randomness as, 6
  statistical, 12, 42
  testing. *See* Hypothesis testing

P, 42

Pattern, 12, 53, 60, 61

Philosophy of science, 46, 58, 117–125
  as metascience, 8
  atomistic vs. organismal science, 58–59
  physics envy, 50
  role of, 7

Platt, J. R. *See* Strong inference

Popper, Karl
  philosophy of, 1, 7, 22–23, 47, 51–52, 58

Positive interactions, 100–102

Power analysis, 38–44
  asserting no difference, 41
  measuring degree, 42–44
  replications, 41
  significance, 42
  sufficient information for, 40
  use to determine sample size, 40

Prediction, 14–15, 50, 58, 122

Proof, 4–5, 48

Protocol, general, 5. *See also* Connell, Joseph, protocol of; Hypothetico-deductive method; Model construction; Inductive method; Popper, Karl, philosophy of; Strong inference; Theory, evaluating

Replication, 49, 109

Sample size. *See* Statistical sample size

Sampling procedure, 11–12. *See also* Statistical sample size

Science. *See* Philosophy of science

Scientific method, 124. *See also* Connell, Joseph, protocol of; Empirical method; Empiricism; Experiment; Experimental design; Hypothetico-deductive method; Model construction; Inductive method; Philosophy of science; Strong inference; Theory, evaluating

Seasons. *See* Weather

Selection pressure, 10

Simplification
  by modeling, 14–17
  vs. perceptiveness, 17

Simultaneous processes. *See* multiple causality

Sparseness (of plants), 72

Species number proportions, 66

Statistical analysis, 38–44, 60–61, 120
  asserting no difference, 41
  chi-square test, 39–40, 43
  contingency test, 43
  effect size, 40, 42
  in everyday life, 4
  measuring degree, 42–44

Statistical hypotheses
  vs. Popperian hypotheses, 12, 23, 38
  vs. strong inference, 25

Statistical sample size, 40, 80

Stochasticity, 60–61
  defined, 60
  dispersal/extinction on islands, 12

Strong inference, 22–26, 32, 34, 47
  defined, 22

Succession models and non-alternative hypotheses, 26–27, 34

Systems dimensions
  and coevolution, 10–11
  and niche theory, 15

Temporal primacy (of hypotheses), 6–7. *See also* Logical primacy

Theory
  antagonism toward, 17
  defined, 14
  developing. *See* Model construction
  evaluating, 3–4, 13, 33, 46–48, 50, 121–124. *See also* Hypothesis testing; Popper, Karl, philosophy of
  purpose, 14, 16–19, 40, 48, 49–51, 62–63, 110

Transplant/removal experiments to assess genetic difference, 9

Truth, 4

Type I error, 13. *See also* Power analysis
  cost vs. type II error, 39–42
  defined, 38

Type II error, 13, 54. *See also* Power analysis
  cost vs. type I error, 39–42
  defined, 38

Univariate tests, 24

Variance, 60

Weather, 61, 64, 65, 68, 70

Zonation of plants, 72